THE PERFECT WEAPON

DAVID E. SANGER is national security correspondent for *The New York Times* and the bestselling author of *The Inheritance* and *Confront and Conceal*. He has been a member of three teams that won the Pulitzer Prize, including in 2017 for international reporting. A regular contributor to CNN, he also teaches national security policy at Harvard's Kennedy School of Government.

THE PERFECT WEAPON

War, Sabotage, and Fear in the Cyber Age_

DAVID E. SANGER

SCRIBE
Melbourne • London

Scribe Publications Pty Ltd
2 John St, Clerkenwell, London, WC1N 2ES, United Kingdom
18–20 Edward St, Brunswick, Victoria 3056, Australia

Published by Scribe 2018

Printed and bound in the UK by CPI Group (UK) Ltd, Croydon CR0 4YY

Scribe Publications is committed to the sustainable use of natural resources and the use of paper products made responsibly from those resources.

9781911617723 (UK edition)
9781925713626 (Australian edition)
9781925693379 (e-book)

CiP records for this title are available from the National Library of Australia and the British Library.

scribepublications.co.uk
scribepublications.com.au

For Sherill,
whose love and talent make all the wonderful things in life happen

CONTENTS

THE PERFECT WEAPON

PREFACE

A YEAR INTO Donald J. Trump's presidency, his defense secretary, Jim Mattis, sent the new commander-in-chief a startling recommendation: with nations around the world threatening to use cyberweapons to bring down America's power grids, cell-phone networks, and water supplies, Trump should declare he was ready to take extraordinary steps to protect the country. If any nation hit America's critical infrastructure with a devastating strike, even a non-nuclear one, it should be forewarned that the United States might reach for a nuclear weapon in response.

Like most things in Washington, the recommendation leaked immediately. Many declared it a crazy idea, and wild overkill. While nations had turned their cyberweapons against each other dozens of times in recent years, no attack had yet been proven to cost a human life, at least directly. Not the American attacks on Iran's and North Korea's weapons programs; not the North Korean attacks on American banks, a famed Hollywood studio, and the British healthcare system; not the Russian attacks on Ukraine, Europe, and then the core of American democracy. That streak of luck was certain to end soon. But why would Donald Trump, or any of his successors, take the huge risk of escalating a cyberwar by going nuclear?

The Pentagon's recommendation, it turned out, was the prelude to other proposals—delivered to a president who values toughness and "America First"—to use the nation's powerful cyberweapons far more aggressively. But it was also a reminder of how quickly the fear of devastating cyberattacks has moved from the stuff of science fiction and *Die Hard* movies to the center of American defense strategy. Just over a decade before, in 2007, cyberattacks were missing entirely from the global "Threat Assessment" that intelligence agencies prepare each year for Congress. Terrorism topped that list, along with other post-9/11 concerns. Now that hierarchy has been reversed: For several years a variety of cyber threats, ranging from a paralyzing strike on the nation's cities to a sophisticated effort to undercut public confidence in its institutions, has appeared as the number one threat on the list. Not since the Soviets tested the Bomb in 1949 had the perception of threats facing the nation been revised so quickly. Yet Mattis, who had risen to four-star status in a career focused on the Middle East, feared that the two decades spent chasing al Qaeda and ISIS around the globe had distracted America from its most potent challenges.

"Great power competition—not terrorism—is now the primary focus of US national security," he said in early 2018. America's "competitive edge has eroded in every domain of warfare," including the newest one, "cyberspace." The nuclear strategy he handed Trump gave voice to an inchoate fear among many in the Pentagon that cyberattacks posed a threat unlike any other, and one we had completely failed to deter.

The irony is that the United States remains the world's stealthiest, most skillful cyberpower, as the Iranians discovered when their centrifuges spun out of control and the North Koreans suspected as their missiles fell out of the sky. But the gap is closing. Cyberweapons are so cheap to develop and so easy to hide that they have proven irresistible. And American officials are discovering that in a world in which almost everything is connected—phones, cars, electrical grids, and satellites—everything can be disrupted, if not destroyed. For seventy years, the thinking inside the Pentagon was that only nations with nuclear weapons could threaten America's existence. Now that assumption is in doubt.

In almost every classified Pentagon scenario for how a future confrontation with Russia and China, even Iran and North Korea, might play out, the adversary's first strike against the United States would include a cyber barrage aimed at civilians. It would fry power grids, stop trains, silence cell phones, and overwhelm the Internet. In the worst-case scenarios, food and water would begin to run out; hospitals would turn people away. Separated from their electronics, and thus their connections, Americans would panic, or turn against one another.

The Pentagon is now planning for this scenario because it knows many of its own war plans open with similarly paralyzing cyberattacks against our adversaries, reflecting new strategies to try to win wars before a shot is fired. Glimpses of what this would look like have leaked out in recent years, partly thanks to Edward J. Snowden, partly because a mysterious group called the Shadow Brokers—suspected of close links to Russian intelligence—obtained terabytes of data containing many of the "tools" that the National Security Agency used to breach foreign computer networks. It didn't take long for some of those stolen cyberweapons to be shot back at America and its allies, in attacks whose bizarre-sounding names, like WannaCry, suddenly appeared in the headlines every week.

Yet the secrecy surrounding these programs obscures most public debate about the wisdom of using them, or the risks inherent in losing control of them. The government's silence about America's new arsenal, and its implications, poses a sharp contrast to the first decades of the nuclear era. The horrific scenes of destruction at Hiroshima and Nagasaki not only seared the national psyche, but they made America's destructive capabilities—and soon Russia's and China's—obvious and undeniable. Yet even while the government kept the details classified—how to build atomic weapons, where they are stored, and who has the authority to order their launch—America engaged in a decades-long political debate about when to threaten to use the Bomb and whether to ban it. Those arguments ended up in a very different place from where they began: in the 1950s the United States talked casually about dropping atomic weapons to end the Korean War; by the eighties there was a national consensus that the US would reach for nuclear weapons only if our national survival was at stake.

So far, there has been no equivalent debate about using cyber-weapons, even as their destructive power becomes more evident each year. The weapons remain invisible, the attacks deniable, the results uncertain. Naturally secretive, intelligence officials and their military counterparts refuse to discuss the scope of America's cyber capabilities for fear of diminishing whatever narrow advantage the country retains over its adversaries.

The result is that the United States makes use of this incredibly powerful new weapon largely in secret, on a case-by-case basis, before we fully understand its consequences. Acts that the United States calls "cyber network exploitations" when conducted by American forces are often called "cyberattacks" when American citizens are the target. That word has come to encompass everything from disabling the grid, to manipulating an election, to worrying about that letter arriving in the mail warning that someone—maybe criminals, maybe the Chinese— just grabbed our credit cards, Social Security numbers, and medical histories, for the second or third time.

During the Cold War, national leaders understood that nuclear weapons had fundamentally changed the dynamics of national secu-rity, even if they disagreed on how to respond to the threat. Yet in the age of digital conflict, few have a handle on how this new revolution is reshaping global power. During his raucous 2016 presidential cam-paign, Trump told me in an interview that America was "so obsolete in cyber," ignoring, if he was aware of it, that the United States and Israel had deployed the most sophisticated cyberweapon in history against Iran. More concerning was the fact that he showed little understanding of the dynamics of the grinding, daily cyber conflict now under way— the short-of-war attacks that have become the new normal. His refusal to acknowledge Russia's pernicious role in the 2016 election, for fear it would undercut his political legitimacy, only exacerbates the problem of formulating a national strategy. But the problem goes far beyond the Trump White House. After a decade of hearings in Congress, there is still little agreement on whether and when cyberstrikes constitute an act of war, an act of terrorism, mere espionage, or cyber-enabled vandalism. Technological change wildly outpaces the ability of politicians— and the citizens who have become the collateral damage in the daily combat of cyberspace—to understand what was happening, much less to devise a national response. Making matters worse, when Russia used social media to increase America's polarization in the 2016 election, the animus between tech companies and the US government—ignited by Snowden's disclosures four years earlier—only deepened. Silicon Valley and Washington are now the equivalent of a divorced couple living on opposite coasts, exchanging snippy text messages.

Trump accepted Mattis's nuclear recommendation without a moment of debate. Meanwhile the Pentagon, sensing Trump's willingness to demonstrate overwhelming American force in cyberspace as in other military arenas, published a new strategy, envisioning an era of constant, low-level cyber conflict in which America's newly minted cyber warriors would go deep behind enemy lines every day, attacking foreign computer servers before threats to the United States could materialize. The idea was classic preemption, updated for the cyber age, to "stop attacks before they penetrate our cyber defenses or impair our military forces." Other proposals suggested the president should no longer have to approve every cyber strike—any more than he would have to approve every drone strike.

In the chaos of the Trump White House, it was unclear how these weapons would be used, or under what rules. But suddenly we are in new territory.

CYBER CONFLICT REMAINS in the gray area between war and peace, an uneasy equilibrium that often seems on the brink of spinning out of control. As the pace of attacks rises, our vulnerability becomes more apparent each day: in the opening months of 2018, the federal government warned utilities that Russian hackers had put "implants" of malware in the nation's nuclear plants and power grid and then, a few weeks later, added that they were infesting the routers that control the networks of small enterprises and even individual homes. In previous years there has been similar evidence about Iranian hackers inside financial institutions and Chinese hackers siphoning off millions of files detailing the most intimate details of the lives of Americans seeking security clearances. But figuring out a proportionate yet effective response has now stymied three American presidents. The problem is made harder by the fact that America's offensive cyber prowess has so outpaced our defense that officials hesitate to strike back.

"That was our problem with the Russians," James Clapper, President Obama's director of national intelligence, told me one winter afternoon at a diner down the road from the CIA headquarters in McLean, Virginia. There were plenty of ideas about how to get back at Putin: unplug Russia from the world's financial system; reveal Putin's links to the oligarchs; make some of his own money—and there was plenty hidden around the world—disappear.

Yet, Clapper noted, "every time someone proposed a way to strike back at Putin for what he was doing in the election, someone else would come back and say, 'What happens next? What if he gets into the voting system?'"

Clapper's question drives to the heart of one of the cyberpower conundrums. The United States can't figure out how to counter Russian attacks without incurring a great risk of escalation. The problem can be paralyzing. Russia's meddling in the election encapsulates the challenge of dealing with this new form of short-of-war aggression. Large and small powers have gradually discovered what a perfect digital weapon looks like. It is as stealthy as it is effective. It leaves opponents uncertain about where the attack came from, and thus where to fire back. And we struggle to figure out the best form of deterrence. Is it better to threaten an overwhelming counterstrike? A non-cyber response, from economic sanctions to using a nuclear weapon? Or to so harden our defenses—a project that would take decades—that enemies give up attacking?

Naturally, the first temptation of Washington policy makers is to compare the problem to something more familiar: defending the country against nuclear weapons. But the nuclear comparison is faulty, and as the cyber expert James Lewis has pointed out, the false analogy has kept us from accurately understanding how cyber plays into the daily geopolitical conflict.

Nuclear arms were designed solely for fighting and winning an overwhelming victory. "Mutually assured destruction" deterred nuclear exchanges because both sides understood they could be utterly destroyed. Cyberweapons, in contrast, come in many subtle shades, ranging from the highly destructive to the psychologically manipulative.

Until recently, Americans were fixated on the most destructive class of cyberweapons, the ones that could turn off a nation's power or interfere with its nuclear command-and-control systems. That is a risk, but the extreme scenario, and perhaps the easier to defend against. Far more common is the daily use of cyberweapons against civilian targets to achieve a more specific mission—neutralizing a petrochemical plant in Saudi Arabia, melting down a steel mill in Germany, paralyzing a city government's computer systems in Atlanta or Kiev, or threatening to manipulate the outcome of elections in the United States, France, or Germany. Such "dialed down" cyberweapons are now used by nations every day, not to destroy an adversary but rather to frustrate it, slow it, undermine its institutions, and leave its citizens angry or confused. And the weapons are almost always employed just below the threshold that would lead to retaliation.

Rob Joyce, Trump's cyber czar for the first fifteen months of the administration and the first occupant of that office to have once run American offensive cyber operations, described in late 2017 why the United States is particularly vulnerable to these kinds of operations, and why our vulnerabilities won't go away anytime soon.

"So much of the fabric of our society rests on the bedrock of our

IT," said Joyce, who spent years running the Tailored Access Operations unit of the NSA, the elite operation charged with breaking into foreign computer networks. "We continue to digitize things; we store our wealth and treasure there; we run operations; we keep our secrets all in that cyber domain." In short, we are inventing new vulnerabilities faster than we are eliminating old ones.

Rarely in human history has a new weapon been adapted with such speed, customized to fit so many different tasks, and exploited by so many nations to reshape their influence on global events without turning to outright war. Among the fastest adapters has been Putin's Russia, which deserves credit as a master of the art form, though it is not the only practitioner. Moscow has shown the world how hybrid war works. The strategy is hardly a state secret: Valery Gerasimov, a Russian general, described the strategy in public, and then helped implement it in Ukraine, the country that has become a test-bed for techniques Russia later used against the United States and its allies. The Gerasimov doctrine combines old and new: Stalinist propaganda, magnified by the power of Twitter and Facebook, and backed up by brute force.

As the story told in this book makes clear, parts of the US government—and many other governments—saw all the signs that our chief adversaries were headed toward a new vector of attack. Yet the United States was remarkably slow to adapt to the new reality. We knew what the Russians had done in Estonia and Georgia a decade ago, the first time they used cyberattacks to help paralyze or confuse an opponent, and we saw what they later attempted from Ukraine to Europe, the testing grounds for cyberweapons of mass disruption and subtle influence. But an absence of imagination kept us from believing that the Russians would dare to leap the Atlantic and apply those same techniques to an election in the United States. And, like the Ukrainians, we took months, even years, to figure out what hit us.

Worse yet, once we began to grasp what happened, a military and intelligence community that prides itself on planning for every contingency had no playbook of ready responses. In early 2018, when asked by the Senate Armed Services Committee how the National Security Agency and US Cyber Command were dealing with the most naked use of cyberpower against American democratic institutions, Adm. Michael S. Rogers, then coming to the end of his term as commander of both organizations, admitted that neither President Obama nor President Trump had given him the authority to respond.

Putin, Rogers said, "has clearly come to the conclusion that there's little price to pay here and that therefore 'I can continue this activity.'" Russia was not alone in reaching this conclusion. Indeed, many adversaries used cyberweapons precisely because they believed them to be a way of undercutting the United States without triggering a direct military response. North Korea paid little price for attacking Sony or robbing central banks. China paid no price for stealing the most private personal details of about 21 million Americans.

The message to our adversaries around the world is clear: cyberweapons, in all their various forms, are uniquely designed to hit America's softest targets. And because they rarely leave smoking ruins, Washington remains befuddled about how to answer all but the biggest and most blatant attacks.

Rogers told me as he began the job in 2014 that his number-one priority was to "establish some cost" for using cyberweapons against America. "If we don't change the dynamic here," he added, "this is going to continue." He left office, in 2018, with the nation facing a far larger problem than when he began.

IN LATE JULY 1909, Wilbur and Orville Wright arrived in Washington to show off their *Military Flyer*. In the grainy pictures that have survived, Washington's swamp creatures streamed across the bridges spanning the Potomac to see the show; even President William Howard Taft got into the act, though the Wright brothers were not about to take the risk of giving him a ride.

Not surprisingly, the army was fascinated by the potential of this wild invention. Generals imagined flying the craft over enemy lines, outflanking an oncoming force, and then sending the cavalry off to dispense with them. It wasn't until three years later, in 1912, that someone thought of arming one of the new "observation aircraft" with a machine gun. Things both ramped up and spiraled down from there. A technology first imagined as a revolutionary means of transportation revolutionized war overnight. In 1913 there were fourteen military airplanes manufactured in the United States; five years later, with World War I raging, there were fourteen thousand.

And they were being used in ways the Wrights never imagined. The Red Baron shot down his first French aircraft in April 1916, over Verdun. Dogfights became monthly, then weekly, then daily events. By World War II, Japanese Zeros were bombing Pearl Harbor and performing kamikaze raids on my father's destroyer in the Pacific. (They missed, twice.) Thirty-six years after Orville's first flights in front of President Taft, the *Enola Gay* banked over Hiroshima and changed the face of warfare forever, combining the reach of airpower and the destructive force of the world's newest ultimate weapon.

In the cyber world today, we are somewhere around World War I. A decade ago there were three or four nations with effective cyber forces; now there are more than thirty. The production curve of weapons produced over the past ten years roughly follows the trajectory of military aircraft. The new weapon has been fired, many times, even if its effects are disputed. As of this writing, in early 2018, the best estimates suggest there have been upward of two hundred known state-on-state cyberattacks over the past decade or so—a figure that describes only those that have become public.

And, as in World War I, this glimpse into the future has led nations to arm up, fast. The United States was among the first, building "Cyber Mission Forces," as they call them—133 teams, totaling more than 6,000 troops, were up and running by the end of 2017. While this book deals largely with the "Seven Sisters" of cyber conflict—the

United States, Russia, China, Britain, Iran, Israel, and North Korea—
nations from Vietnam to Mexico are emulating the effort. Many have
started at home by testing their cyber capabilities against dissidents and
political challengers. But no modern military can live without cyber
capabilities, just as no nation could imagine, after 1918, living without
airpower. And now, as then, it is impossible to imagine fully how dramatically this invention will alter the exercise of national power.

In 1957, with the world on the nuclear precipice, a young Harvard
scholar named Henry Kissinger wrote *Nuclear Weapons and Foreign
Policy.* The book was an effort to explain to an anxious American public how the first use, a dozen years before, of a powerful new weapon
whose implications we barely understood was fundamentally reorder-
ing power around the world.

One doesn't have to endorse Kissinger's conclusions in that book—
especially his suggestion that the United States could fight and survive
a limited nuclear war—to admire his understanding that after the invention of the Bomb, nothing could ever be the same. "A revolution
cannot be mastered until it is understood," he wrote. "The temptation
is always to seek to integrate it into familiar doctrine: to deny a revolution is taking place." It was time, he said, "to attempt an assessment
of the technological revolution which we have witnessed in the past
decade" and to understand how it affected everything we once thought
we understood. The Cuban Missile Crisis erupted only five years later,
the closest the world came in the Cold War to annihilation by miscalculation. That crisis was followed by the first efforts to control the
spread of nuclear weapons before they dictated our fate.

While most nuclear analogies do not translate well to the new world
of cyber conflict, this one does: We all live in a state of fear of how
our digital dependencies can be hijacked by nations that in the past
decade have discovered a new way to pursue old struggles. We have
learned that cyberweapons, like nuclear weapons, are a great leveler.

And we worry, with good reason, that within just a few years these weapons, merged with artificial intelligence, will act with such hyperspeed that escalatory attacks will take place before humans have the time—or good sense—to intervene. We keep digging for new technological solutions—bigger firewalls, better passwords, better detection systems—to build the equivalent of France's Maginot Line. Adversaries do what Germany did: they keep finding ways around the wall.

Great powers and once-great powers, like China and Russia, are already thinking forward to a new era in which such walls pose no obstacle and cyber is used to win conflicts before they appear to start. They look at quantum computers and see a technology that could break any form of encryption and perhaps get into the command-and-control systems of America's nuclear arsenal. They look at bots that could not only replicate real people on Twitter but paralyze early-warning satellites. From the NSA headquarters at Fort Meade to the national laboratories that once created the atomic bomb, American scientists and engineers are struggling to maintain a lead. The challenge is to think about how to defend a civilian infrastructure that the United States government does not control, and private networks where companies and American citizens often don't want their government lurking—even for the purpose of defending them.

What's missing in these debates, at least so far, is any serious effort to design a geopolitical solution in addition to a technological one. In my national security reporting for the *New York Times*, I've often been struck by the absence of the kind of grand strategic debates surrounding cyber that dominated the first nuclear age. Partly that is because there are so many more players than there were during the Cold War. Partly it is because the United States is so politically divided. Partly it is because cyberweapons were created by the US intelligence apparatus, instinctively secretive institutions that always err on the side of overclassification and often argue that public discussion of how we might want to use or control these weapons imperils their utility.

Some of that secrecy is understandable. Vulnerabilities in computers and networks—the kind that allowed the United States to slow

Iran's nuclear progress, peer inside North Korea, and trace Russia's role in the 2016 election—are fleeting. But there is a price for secrecy, and the United States has begun to pay that price. It is impossible to begin to negotiate norms of behavior in cyberspace until we too are willing to declare our capabilities and live within some limits. The United States, for example, would never support rules that banned cyber espionage. But it has also resisted rules prohibiting the placement of "implants" in foreign computer networks, which we also use in case the United States needs a way to bring those networks down. Yet we are horrified when we find Russian or Chinese implants in our power grid or our cell-phone systems.

"The key issue, in my opinion," says Jack Goldsmith, a Harvard law professor who served in George W. Bush's Justice Department, "is the US government's failure to look in the mirror."

ON A SUMMER day in 2017, I went to Connecticut to see Kissinger, who was then ninety-four, and asked him how this new age compared to what he grappled with in the Cold War. "It is far more complex," he said. "And over the long-term, it may be far more dangerous."

This book tells the story of how that complexity and danger are already reshaping our world, and explores whether we can remain masters of our own invention.

FROM RUSSIA, WITH LOVE

A S THE LIGHTS went out in western Ukraine the day before Christmas Eve 2015, Andy Ozment had a queasy feeling.

The giant screens in the war room just down the hall from his office—in an unmarked Department of Homeland Security building a quick drive over the Potomac River from the White House—indicated that something more nefarious than a winter storm or a blown-up substation had triggered the sudden darkness across a remote corner of the embattled former Soviet republic. The event had all the markings of a sophisticated cyberattack, remote-controlled from someplace far from Ukraine.

It had been less than two years since Vladimir V. Putin had annexed Crimea and declared it would once again be part of Mother Russia. Putin's tanks and troops—who traded in their uniforms for civilian clothing and became known as the "little green men"—were sowing chaos in the Russian-speaking southeast of Ukraine, and doing what they could to destabilize a new, pro-Western government in Kiev, the capital.

Ozment knew that a Russian cyberattack against Ukrainians, far from the active combat zones, would make sense now, in the middle of the holidays. The electric utility providers were operating with skeleton staffs. To Putin's secret army of patriotic hackers, Ukraine was a playground and testing ground. What happened there, Ozment often told his staff, was a prelude to what might well happen in the United States. As he regularly reminded them, in the world of cyber conflict, attackers came in five distinct varieties: "vandals, burglars, thugs, spies, and saboteurs."

"I'm not that worried about the thugs, the vandals, and the burglars," he would quickly add. It was up to companies and government agencies to guard against the run-of-the-mill bad actors on the Internet. It was the spies—and particularly the saboteurs—who kept him up at night. And the saboteurs who hit Ukraine's power grid in 2015 were not amateurs. "All the advantages go to the attacker," Ozment warned. Putin appeared to be making that point in Ukraine.

A bearded computer scientist in his late thirties, Ozment seemed to deliberately cultivate a demeanor suggesting it hadn't been that long since he graduated from Georgia Tech and that he'd rather be hiking than cracking malware. He lived with his Norwegian wife in a two-story redbrick townhouse in a funky section of Washington, north of the Capitol. He always managed to look like he just walked out of one of the weekend farmers markets in his neighborhood, rather than off the front lines of America's daily cyberwars. It was an admirable feat, considering he was running the closest thing the US government had to a fire department for cyberattacks. His team in Arlington functioned as the first responders when banks or insurance companies were attacked, utility companies found viruses lurking in their networks and suspected foul play, or incompetent federal agencies—like the Office of Personnel Management—discovered that Chinese intelligence agents were walking off with millions of highly sensitive security-clearance files. In other words, Ozment's team got called all the time, like an engine company in a neighborhood of arsonists.

Ozment's cyberwar room—in bureaucratese, the "National Cybersecurity & Communications Integration Center"—looked like a Hollywood set. The screens ran for more than a hundred feet, showing everything from the state of Internet traffic to the operation of power plants. Tickers with news items sped by. The desks in front of the screens were manned by various three-letter agencies in the US government: the Federal Bureau of Investigation, the Central Intelligence Agency, the National Security Agency, the Department of Energy.

At first glance, the room resembled the kind of underground bunker that a previous generation of Americans had manned round the clock, in a mountain near Colorado Springs. But initial impressions were deceiving. The men and women who spent the Cold War glued to their giant screens in Colorado were looking for something that was hard to miss: evidence of nuclear missiles speeding into space, aimed at American cities and silos. If they saw a launch—and there were many false alarms—they knew they had only minutes to confirm the US was under attack and to provide warning to the president, who would have to decide whether to retaliate before the first blast. But there was a certain clarity: At least they could know who launched the missiles, where they came from, and how to retaliate. That clarity created a framework for deterrence.

Ozment's screens, by contrast, provided proof that in the digital age, deterrence stops at the keyboard. The chaos of the modern Internet played out across screen after screen, often in an incomprehensible jumble. There were innocent service outages and outrageous attacks, yet it was almost impossible to see where any given attack came from. Spoofing the system came naturally to hackers, and masking their location was pretty simple. Even in the case of a big attack, it would take weeks, or months, before a formal intelligence "attribution" would emerge from American intelligence agencies, and even then there might be no certainty about who had instigated the attack. In short, it was nothing like the nuclear age. Analysts could warn the president about what was happening—and Ozment's team often did—but they could not specify, in real time and with certainty, where an attack was coming from or against whom to retaliate.

The more data that flowed in about what was happening that winter day in Ukraine, the deeper Ozment's stomach sank. "This was the kind of nightmare we've talked about and tried to head off for years," he recalled later. It was a holiday week, a rare break from the daily string of crises, and Ozment had a few minutes to dwell on a chilling cell-phone video that his colleagues were passing around. Taken in the midst of the Ukraine attack by one of the operators at the beleaguered electricity provider, Kyivoblenergo, it captured the bewilderment and chaos among electric-grid operators as they frantically tried to regain control of their computer systems.

As the video showed, they were helpless. Nothing they clicked had any effect. It was as if their own keyboards and mice were disconnected, and paranormal powers had taken over their controls. Cursors began jumping across the screens at the master control center in Ukraine, driven by a hidden hand. By remote control, the attackers systematically disconnected circuits, deleted backup systems, and shut down substations. Neighborhood by neighborhood, the lights clicked off. "It was jaw-dropping for us," said Ozment. "The exact scenario we were worried about wasn't paranoia. It was playing out before our eyes."

And the hackers had more in store. They had planted a cheap program—malware named "KillDisk"—to wipe out the systems that would otherwise allow the operators to regain control. Then the hackers delivered their finishing touch: they disconnected the backup electrical system in the control room, so that not only were the operators now helpless but they were sitting in darkness. All the Kyivoblenergo workers could do was sit there and curse.

For two decades—since before Ozment began his career in cyber defense—experts had warned that hackers might switch off a nation's power grid, the first step in taking down an entire country. And for most of that time, everyone seemed certain that when the big strike came, it would take out the power from Boston to Washington, or San Francisco to Los Angeles. "For twenty years we were paranoid about it, but it had never happened," Ozment recalled.

"Now," he said, "it was happening."

．　．　．

IT WAS HAPPENING, but on a much broader scale, in ways that Ozment could not yet imagine.

While Ozment struggled to understand the implications of the cyberattack unfolding half a world away in Ukraine, the Russians were already deep into a three-pronged cyberattack on the very ground beneath his feet. The first phase had targeted American nuclear power plants as well as water and electric systems, with the insertion of malicious code that would give Russia the opportunity to sabotage the plants or shut them off at will. The second was focused on the Democratic National Committee, an early victim in a series of escalating attacks ordered, American intelligence agencies later concluded, by Vladimir V. Putin himself. And the third was aimed at the heart of American innovation, Silicon Valley. For a decade the executives of Facebook, Apple, and Google were convinced that the technology that made them billions of dollars would also hasten the spread of democracy around the world. Putin was out to disprove that thesis and show that he could use those same tools to break democracy and enhance his own power.

It added up to a multifaceted attack on America's infrastructure and institutions, and was remarkable in its scope, startling in its brazenness. Americans were shocked, but Putin's moves had hardly come out of the blue. They were merely the latest phase of a global battle fought over unseen networks for the better part of a decade—a battle in which America had fired some of the opening shots.

CHAPTER I

ORIGINAL SINS

This has a whiff of August 1945. Somebody just used a new weapon, and this weapon will not be put back in the box.
—*Gen. Michael Hayden, former director, National Security Agency and Central Intelligence Agency*

ON AN EARLY spring day in 2012, I drove along the winding, wooded driveway of the Central Intelligence Agency and pulled up in front of what the agency quaintly calls its "Old Headquarters."

I knew that the meeting I was headed to—with Michael Morell, the agency's deputy director—was likely to be difficult. A few weeks before, the White House had asked me to see Morell and talk with him about an especially sensitive story the *Times* was preparing to publish. The two of us had met briefly in the West Wing basement office of Benjamin J. Rhodes, then the deputy national security advisor for strategic communications, as I explained what I had learned: how two presidents of strikingly different temperaments, George W. Bush and Barack Obama, had both come to the decision to use the most sophisticated cyberweapon in history against Iran as the last, best chance to forestall a new war in the Middle East.

Neither Rhodes nor Morell seemed surprised that I had pieced the story together; the weapon's code, called "Stuxnet," had accidentally spread around the world nearly two years before, making it evident that someone was using malware in an attempt to blow up Iran's nuclear facilities. Stuxnet was filled with digital fingerprints and other clues about where and when it had been written. That someone eventually would follow those clues to discover the plan that had launched it seemed inevitable. The operation, which I learned through months of reporting had been code-named "Olympic Games," was simply too big, and involved too many players, to stay secret forever. The Iranians themselves had long ago declared, with relatively little proof, that the United States and Israel were behind the attack. But neither government had ever uttered an admission, emblematic of the reflexive secrecy they wrapped around all cyber operations.

As is true with nuclear weapons, only the president could authorize the American use of a cyberweapon for destructive purposes. Yet because virtually all offensive cyber operations take place as covert actions, which by law must be designed to be deniable, no American president had ever been caught authorizing one. The *Times* piece would lay out the Situation Room debate over using a cyberweapon to mount the kind of attack that, previously, could have been executed only by bombing or sending in saboteurs.

But as I walked across the famous atrium of the CIA—its walls dotted with bronze-colored stars, one for each of the CIA officers who had died in defense of the country—and headed up the elevator to Morell's office, there was no way for me to know how the story threatened to disrupt the web of secrecy the United States had built around its decade-long race to build up its cyber capabilities. Nor could I have known that I would touch off one of the larger federal leak investigations of modern times, or that it would lead to the unfair prosecution of a military officer who was highly valued by Obama and who was among the cohort who had brought the US military into the era of modern cyberwarfare.

It turned out that the US government was not yet ready to discuss the consequences of its decision to use cyberweapons against another state in peacetime. Nor was it eager to assess the degree to which

America's actions contributed to a cyber arms race that Iran, Russia, North Korea, and China had joined.

BEYOND THE OFT-PHOTOGRAPHED lobby, the CIA's well-worn executive offices resembled those of the declining computer firms—like the now-extinct Burroughs and Digital Equipment Corporation—that I had covered decades ago as a young technology reporter. The retro look prevailed especially on the seventh floor, in the suite that Allen Dulles, CIA director under Eisenhower and Kennedy, designed so that he could sit within feet of his deputy director as they oversaw the vast and complex Cold War effort to steal secrets and take down adversaries. Appropriately, the look of the world's most famous spy agency was a bit deceiving: As the story of Olympic Games made clear, the agency was deeply into the digital age. But it had no interest in overtly displaying its prowess.

I had come to the Old Headquarters to hear about which details of the emerging story so concerned Morell and his colleagues that they were preparing to ask the *Times* to withhold them, lest we tip off other targets of ongoing operations. By their nature such conversations are fraught. News organizations must be willing to listen to government concerns, but insist, for obvious First Amendment reasons, that the decision to publish belongs to them, not the government. Morell, while always friendly and professional, had already indicated that, in his view, none of the Olympic Games story should be published. But he was a realist, and knew that the accidental revelation and dissemination of the Stuxnet worm meant the story was not going to disappear. For the CIA, that day's meeting was an exercise in learning what I had learned, and in directing damage control.

Operation Olympic Games was largely the work of the NSA and Israel's Unit 8200, its military cyber operation. But the CIA, I had learned over time, played a key part, executing a presidential authorization for covert action—known in Washington as a "finding"—to slow Iran's nuclear program. Because "findings" are secret and intended to be denied publicly, I had no expectation that the agency officials I saw that day would acknowledge their role in deploying the weapon, much less the subsequent destruction of roughly one thousand centrifuges that had been spinning beneath the Iranian desert. And they did not.

But something about this story was different, and it added to the tension over its forthcoming publication. Cyberweapons, among the first strategic weapons created by the intelligence agencies rather than by the military, had been swaddled in more secrecy than that surrounding nuclear and biological weapons, or new generations of stealth fighters and drones. There was an assumption inside the government that anything published about the use of cyberweapons would impede their future use. While the government would describe in great detail its outrage about cyberattacks against the United States—or even trace evidence that other powers had entered the networks of our banks or electric systems—it considered the most basic conversations about US capabilities, intentions, or doctrines off-limits. Even some inside the US government deemed this level of secrecy ridiculous: How could you begin to discuss setting international rules about the use of weapons you won't acknowledge owning, much less using?

Clearly, there was no consensus within the Obama administration about how these weapons should be used. Even while Obama was approving new strikes on the Iranian nuclear plant, he harbored his own doubts. As our story explained, in meetings in the Situation Room in the first year of his presidency, Obama had repeatedly questioned whether the United States was setting a precedent—using a cyberweapon to cripple a nuclear facility—that the country would one day regret. This was, he and others noted, exactly the kind of precision-guided weapon that other nations would someday learn to turn on us. "It was the right question," said one senior official who came into the administration after the Stuxnet attacks were over. "But no one understood how quickly that day would come."

Curiously, Obama had already proven willing to engage in a public argument over similar questions about drones. Everything about drone warfare had been secret when he came to office, but over time Obama made elements of the program public and proved willing to explain the law and reasoning behind his decision to deploy these remote-controlled killing machines. In doing so, he gradually lifted the secrecy surrounding the use of drones so that the world could understand whether they were hitting terrorists, and when they went awry and killed children or wedding guests.

Cyberweapons were different. The government would barely admit to owning them, much less talk about the rules for when and why it used them. But the issues were very similar; just as investigative reporting about the unintended costs of drone strikes had forced the debate about unmanned weapons, my editors and I felt a journalistic imperative to explain to readers how the government was embracing cyberweapons that could ultimately be turned against our homeland. Olympic Games had opened the door to a new dimension of warfare that no one fully understood.

The only thing that was clear was that there would be no backpedaling. When Michael Hayden, who had been central to the early days of America's experimentation with cyberweapons, said that the Stuxnet code had "the whiff of August 1945" about it—a reference to the dropping of the atomic bomb on Hiroshima and Nagasaki—he was making clear that a new era had dawned. Hayden's security clearances meant he couldn't acknowledge American involvement in Stuxnet, but he left no doubt about the magnitude of its importance.

"I do know this," Hayden concluded. "If we go out and do something, most of the rest of the world now feels that this is a new standard, and it's something that they now feel legitimated to do as well."

That is exactly what happened.

. . .

HAYDEN WAS WELL practiced at talking about Stuxnet as if he were an outsider looking in, a zoologist who had just observed the odd behavior of an animal and declared the discovery of a new species. But in fact, he likely knew exactly what he was looking at. Hayden served as director of the CIA during the early days of Olympic Games. By that time, he was already in the vanguard of those who, in the mid-1990s, came to believe that cyberweapons were not simply a new tool but also what war fighters call a "new domain": the place where future power conflicts great and small would play out.

As Hayden rose through the ranks of the air force in the 1970s, everyone agreed on the four physical domains that had long defined warfare: People had fought on land and sea for millennia, and in the air since World War I. Space was added in the 1950s and '60s, when satellites begat antisatellites, and intercontinental ballistic missiles led to antiballistic missile systems. But cyberspace? As one long-retired general once asked me with genuine mystification at the Air Force Academy in Colorado Springs, "How do you fight in a place you can't see?"

Hayden's insight into the game-changing nature of cyber conflict began to form more than twenty years earlier, when he was assigned to San Antonio, Texas, as the commander of the Air Intelligence Agency, an air force unit that gave him an early glimpse of the power of a new generation of electronic weapons. He remembered watching in wonder as members of the staff disabled remote workstations and used electronic-warfare techniques to fool a radarscope that was trying to track a fighter plane. But what struck him most, fresh back from the wars in the Balkans, was how relentlessly the US military was coming under regular attack in peacetime.

The year after Hayden got to Texas, in 1998, the FBI was called in to investigate seemingly bizarre intrusions that had begun popping up in strange places connected to military or intelligence networks, from the Los Alamos and Sandia National Laboratories—where nuclear weapons are designed—to universities, such as the Colorado School of Mines, which held a significant contract with the Navy. There was a particular concentration of intrusions around the networks of Wright-Patterson Air Force Base in Ohio, located on the site where the Wright brothers once tested many of their early planes.

It was a computer operator at the School of Mines who first discovered the hack, after he saw some nighttime computer activity he could not explain. The attack turned out to be a very large one, and persistent, seemingly coming from Russia. The hackers had lurked in some of these systems for two years and had stolen thousands of pages of unclassified material concerning sensitive technologies.

Shock soon gave way to the accompanying recognition of a new reality. The attack was given the name "Moonlight Maze." The Russians were initially helpful in the investigation, until they realized that the FBI had evidence that it was the Russian government, not some teenage hackers, behind the intrusions. Moscow shut down its cooperation. John Hamre, the bookish, usually unflappable defense scholar who was serving as deputy secretary of defense, told Congressional intelligence committees, "We're in the middle of a cyberwar."

"This was a real wake-up call for us," Hamre told me. "Until then, we'd had incursions, but never a case of a foreign power that broke into our systems and simply wouldn't leave—and was hard to evict."

Some experts who have studied the intrusion argue that Moonlight Maze never really ended; it just morphed into new attacks that continued for the next two decades. Whatever the truth, the Russian attacks galvanized the first serious efforts by the United States to defend its networks and form its own offensive cyber forces.

The attack forced the United States to confront the implications of the digital age. As Hayden noted, in the 1980s, when he was based in Korea, a military communication would be typed, scanned, sent to Washington, and then printed for someone to deal with as if it were just another piece of classified paper. But suddenly emails and classified cables became the default mode of communication and gave skilled intelligence agencies worldwide a way to intercept a far wider range of information "in transit."

The explosion of digital data gave the NSA a new mission. The agency responsible for encrypting and protecting sensitive information, mostly for the military and intelligence agencies, zeroed in on a vast new set of targets: computer data stored around the world that was vulnerable to the NSA's fast-growing cadres of hackers. Much of this information was not the kind of "data in transit" the NSA had spent decades intercepting. Instead, it was locked away in computer complexes that foreign governments, in their naïveté, had viewed as largely invulnerable. That was a fantasy, of course. An agency that had spent decades intercepting electrons flying through phone lines and over satellites was suddenly focused on what they called "data at rest." And getting that data meant breaking into computer networks around the world.

"This was all about going to the end point, the targeted network," Hayden later wrote, rather than waiting in hopes a message could be plucked out of the sky. And that required figuring out how to break into systems. Soon the NSA, CIA, and Pentagon joined forces to create an organization, blandly called the Information Operations Technology Center, designed to do just that.

The center was regarded with enormous suspicion by old-timers at the CIA who thought it represented game playing by people who should be doing real spying. But these veterans were living in a lost world. In retrospect, by the early 2000s the United States was entering a new arms race, akin to the one in which it had invested billions for the first hydrogen bomb, then the first intercontinental ballistic missiles, and later still missiles with multiple warheads. But even the Pentagon didn't know how to think about these new weapons or where to put them in its vast bureaucracy. Donald Rumsfeld, returning in 2001 to the post of secretary of defense, which he had held in the late 1970s, began searching for a place to house this strange new capability—offensive cyberweapons—in the vastness of the military's combat commands.

From Rumsfeld's recently declassified "snowflakes"—his brief messages to his staff ordering up studies—it is clear that he sensed that cyberweapons were enormously powerful tools. But he struggled to understand how the Pentagon would use them. Naturally, the military had already developed jargon for the variety of techniques, vulnerabilities, and weapons in the arsenal. There were "computer network exploitation" operations, a fancy way of describing the theft of an adversary's data. And there were "computer network attacks," which are cyberattacks with real-world effects of the kind that were later tested in Olympic Games.

"Everything at the Pentagon needs a home," Hamre told me. "And Rumsfeld, rightly, saw this as a strategic weapon and gave it to Hoss Cartwright at Strategic Command."

Gen. James Cartwright, a marine aviator whose nickname, "Hoss," was taken from a character in the '60s television show *Bonanza,* ranked among the best strategic minds in a military consumed by the day-to-day battles in Iraq and Afghanistan. He walked around Strategic Command with a low-key demeanor and a crinkly smile, an everyman look from his days growing up in Rockford, Illinois. Cartwright had been pre-med and a competitive swimmer at the University of Iowa and, in the last days of the Vietnam War, signed up with the marines as a naval aviator. There's no room for error when taking off from and landing on aircraft carriers, and those high stakes appealed to Cartwright's sense of precision. But he also learned that naval aviators can never look like they are sweating the details, even when there is only one chance to catch the cable that keeps a plane from plunging into the sea during a deck landing.

By the time Bush took office in 2001, Cartwright had developed a fascination with the promise, and the danger, of cyberweapons. In his quiet but intense fashion, he began questioning whether the systems and strategies the Pentagon had built up in the decades after World War II were sufficient to meet the challenges of the next fifty years. The answer seemed obvious to him.

Yet inside the Pentagon one could make a lot of enemies questioning whether the conventional weapons that had gotten us through Vietnam and two wars in the Gulf remained critical in an age in which breaking into an industrial control network might be more important than fielding new tanks and bombers. "There were a lot of people in the Pentagon who found Hoss's questioning refreshing," one of the members of the Joint Chiefs who served with him said to me. "And there were a lot who found it threatening."

That was especially true as Cartwright took on his first major job as a marine general in 2004: head of the US Strategic Command in Omaha, Nebraska. There was no job where precision and a strategic view of the world mattered more. Strategic Command, known as Stratcom, is in charge of the nation's nuclear arsenal. During the Cold War, it was the first line of defense against a nuclear conflict with the Soviets and was responsible for maintaining and moving nuclear weapons, drilling its staff for every scenario under which they might be launched, and making sure that any order to use them was authentic and legal. The opportunities for error on a horrific scale were endless.

Cartwright looked at Strategic Command's arsenal and began to ask a big question: Are these really the weapons that will keep the nation safe in the next half century? There were safety issues: the nuclear arsenal was aging; missile silos were still using five-inch floppy disks. The missileers working inside the silos were dispirited; not only were their command posts damp and out of date, but staffers were running through mind-deadening procedures preparing for an order that would probably never come.

Cartwright was equally concerned about the strategic vacuum. America's reliance on nuclear deterrence was actually restricting a president's ability to deal with the kind of adversaries the United States was facing every day, from the Middle East to East Asia. Because the consequences and casualties of using a nuclear weapon were so huge that they were paralyzing, Cartwright began to think strategically about the new cyberweapons that Rumsfeld had put under his command. They presented a huge intellectual puzzle and, as Hayden remembered later, "Hoss was strangely underemployed at Stratcom." He began thinking about how cyberweapons could expand a president's choices after decades in which nuclear weapons had limited them.

"The tools available to a president or nation in between diplomacy and military power were not terribly effective," Cartwright told the US Naval Institute in 2012. He had by then left military service and was only beginning to unspool his thinking on this problem. What American presidents needed, he believed, were more coercive tools that could back up diplomacy. And nuclear weapons did not serve that purpose. No adversary thought an American president would ever reach for a nuclear weapon, except if the survival of the United States were at stake.

In his years at Strategic Command, Cartwright later said, he kept looking for new technologies the military could actually employ and, preferably, exploit so that the United States could prevail in a fight without ever firing a shot. These cyberweapons were what he called "speed-of-light" weapons—repurposed "electronic warfare" weapons that could disable an adversary's communications or paralyze its defenses. Others were directed energy weapons, such as lasers. Unlike nuclear weapons, these could be used in a first strike.

More important, beyond the damage they could inflict in wartime, cyberweapons had a coercive power in peacetime. Cartwright talked about using these weapons "to reset diplomacy," or to force a country to realize that it had little choice other than agreeing to negotiate. When he gave his 2012 speech, Cartwright never once made reference to Iran, but he didn't need to do so. To anyone watching the world scene at the time—a moment when the United States was simultaneously preparing to negotiate with Tehran and to go to war with it—his meaning was obvious.

Soon after Rumsfeld handed cyberwarfare to Strategic Command, a skunk works of sorts popped up there, exploring what it would take to deploy these weapons, how they should be used, and how the military's role in marshaling them would be different from the NSA's role. Over time, what emerged from Cartwright's creation was a prototype of what today is the US Cyber Command, although then it existed largely on paper and was barely staffed.

In 2007, with wars still raging in the Middle East and South Asia,

Cartwright moved on to become vice chairman of the Joint Chiefs of Staff. It was a rough transition. He wasn't an Iraq veteran, a liability at a time that this distinction was cherished as a prerequisite for higher command. Tension developed between Cartwright and the chairman of the Joint Chiefs, Adm. Mike Mullen, and worsened over time. Despite challenges, it was from this post that Cartwright began to put America's cyber forces in action.

IN JANUARY OF that same year, 2007, the director of national intelligence, John D. Negroponte, presented Congress with the annual worldwide threat assessment, an exercise that the nation's top intelligence officials understandably despised. It forced them to rank—in public—the major threats to the United States, and often it was only an exercise in telling Congress what it wanted to hear. But as a snapshot of national fears and obsessions at any given moment, it was nonetheless revealing.

When Negroponte settled into the witness chair that January day, he opened with a blunt statement: "Terrorism remains the preeminent threat to the homeland." Senators nodded in agreement. Dig further into his report, however, and one fact leaped out: cyberattacks did not even make the list. They were totally absent.

Yet even then, the nation's intelligence chiefs knew well that the daily skirmishing among superpowers was, if anything, intensifying. Chinese attacks on American companies—including military contractors— were ramping up. By 2008, the year after Negroponte testified, Chinese hackers working for the People's Liberation Army were inside Lockheed Martin's networks, making off with plans related to the F-35, the world's most sophisticated, and certainly most expensive, fighter jet. Later that year they hacked the campaigns of Barack Obama and John McCain, rivals for the presidency. Lisa Monaco, who was running the national security division of the Justice Department at the time, remembers clearly the first time she met Obama's senior staff. "I went out to explain to them that the Chinese were all over their system," she said with a laugh years later, when she was the Homeland Security Advisor at the White House and overseeing the effort to bolster the nation's cyber defenses.

But the true wake-up call came on October 24, 2008, with the nation on the brink of Obama's election. Debora Plunkett remembers it well. A month into a new job running the NSA's Advanced Network Operations division, she was assigned to develop and deploy tools to determine if anyone was inside, or trying to get inside, the US government's classified networks.

Plunkett hadn't taken a conventional route to the NSA. The daughter of a long-distance trucker, she had grown up not far from Fort Meade but had never heard of the agency until after college. Coming off two tough years in forensics with the Baltimore Police Department, she was advised by a friend's boyfriend who worked for the NSA to take the entrance exam. She was given only a vague description of the agency's work, but for Plunkett, who loved puzzles, what she heard sounded intriguing. She passed the exam and joined the NSA in 1984.

Over the next quarter century, Plunkett became one of the few African American women to rise within the NSA leadership. "I was quite often the only minority and absolutely the only minority woman in my workspace and organization," she said. She climbed from the cryptography section to her position running the ANO and soon found herself leading a search for network intruders.

On a brisk fall day at Fort Meade in 2008—just ahead of Obama's election—Plunkett's team found something that made her blood run cold: Russian intruders in the Pentagon's classified networks. This was a new encroachment for the defense department, which had never—until that moment—discovered a breach in what was known as SIPRNet (it had the unwieldy name of "Secret Internet Protocol Router Network"). SIPRNet was far more than an internal network: It connected the military, senior officials in the White House, and the intelligence agencies.

In short, if the Russians were in that communication channel, they had access to everything that mattered. Plunkett recalls that "pretty soon we went straight to Alexander," meaning Gen. Keith Alexander, then the director of the NSA.

Investigators raced to figure out how the Russians had gotten inside. The answer was pretty shocking: The Russians had left USB drives littered around the parking and public areas of a US base in the Middle East. Someone picked one up, and when they put the drive in a laptop connected to SIPRNet, the Russians were inside. By the time Plunkett and her team made their discovery, the bug had spread to all of US Central Command and beyond and begun scooping up data, copying it, and sending it back to the Russians.

It was a bitter lesson for the Pentagon—they were, in fact, easy pickings for attackers using a technique that the CIA and NSA had often used to get into foreign computer systems. "People worked through the night to come up with a solution," Plunkett recalled. "We were able to develop what we thought was a reasonable solution that ended up being a very good solution." The fix—called Operation Buckshot Yankee— was deployed by the Pentagon later that day. Then, to keep a similar breach from happening again, USB ports on Department of Defense computers were sealed with superglue.

But the damage had already been done. As William Lynn, then deputy secretary of defense, later explained, the intrusion "was the most significant breach of U.S. military computers ever, and it served as an important wake-up call."

Perhaps so, but not everybody woke up. After leaving the NSA, Plunkett told me that for all her efforts—and they were considerable— she remained amazed by how easily outsiders appeared able to break into government and corporate systems. With every major hack, "folks like me will say—this will be the moment, this is the watershed moment. And it never was," she added, "because we're so lax about security and so inconsistent in investing in security.

"We just make it easy for them."

.   .   .

WHILE PLUNKETT WAS trying to fortify the Pentagon's networks against the Russians, the NSA's offensive team, working not far away on the Fort Meade campus, was already making centrifuges blow up in Natanz.

Prodded by General Cartwright, Keith Alexander at the NSA, and a range of other intelligence officials, President Bush had authorized a covert effort to inject malicious code into the computer controllers at the underground Iranian plant. Part of the plan was to slow the Iranians and force them to the bargaining table. But an equally important motivation was to dissuade Prime Minister Benjamin Netanyahu of Israel from bombing Iran's facilities, a threat he was making every few months. Bush took the threat very seriously. Twice before the Israelis had seen threatening nuclear projects under way, one in Iraq, the other in Syria. They had destroyed them both.

Olympic Games was a way to keep the Israelis focused on crippling the Iranian program without setting off a regional war. But getting the code into the plant was no easy task. The Natanz computer systems were "air gapped" from the outside, meaning they had no connections to the Internet. The CIA and the Israelis endeavored to slip the code in on USB keys, among other techniques, with the help of both unwitting and witting Iranian engineers. With some hitches, the plan worked reasonably well for several years. The Iranians were mystified about why some of their centrifuges were speeding up or slowing down and ultimately destroying themselves. Spooked, they pulled other centrifuges out of operation before those met the same fate. They started firing engineers.

At Fort Meade, and the White House, the subterfuge seemed successful beyond anything its creators had hoped. And then all went wrong.

No reporter or news organization exposed Olympic Games. The governments of the United States and Israel managed to do so all by themselves, by mistake. There has since been a lot of finger-pointing about who was responsible, with the Israelis claiming the United States moved too slowly, and the United States claiming the Israelis became impatient and sloppy. But one fact is indisputable: the Stuxnet worm got out into the wild in the summer of 2010 and quickly replicated itself in computer systems around the world.

It showed up in computer networks from Iran to India, and eventually even wound its way back to the United States. Suddenly everyone had a copy of it—the Iranians and the Russians, the Chinese and the North Koreans, and hackers around the globe. That is when it was given the name "Stuxnet," a blend of keywords drawn from inside the code.

In retrospect, Operation Olympic Games was the opening salvo in modern cyber conflict. But at the time, no one knew that. All that could be said for sure was that a strange computer worm floating around the world had emanated from Iran, and in that summer of 2010 Iran's nuclear program seemed a natural target.

In the newsroom of the *Times,* we had been on high alert for any evidence that a cyberweapon, rather than bombs and missiles, was being aimed at Iran's nuclear complex. In early 2009, just as Obama was preparing to take office, I reported that President Bush had secretly authorized a covert plan to undermine electrical systems, computer systems, and other networks on which Iran relies, in the hopes of delaying the day that Iran could produce a workable nuclear weapon. Eighteen months later, no one was surprised when evidence began to mount that Stuxnet was the code we had been looking for.

Soon an unbeatable team of cyber sleuths—Liam O'Murchu and Eric Chien of Symantec—grew intrigued. They were the odd couple of cyber defense: O'Murchu a boisterous Irishman with a thick brogue who raised the alarm at Symantec, and Chien the quiet engineer who dug in. For weeks the pair ground away at the code. They ran it through filters, compared it to other malware, and mapped how it worked. "It's twenty times the size of the average piece of code," but contained almost no bugs, Chien recalled later. "That's extremely rare. Malicious code always has bugs inside of it. This wasn't the case with Stuxnet." He admired the malware as if he were an art collector who had just discovered a never-before-seen Rembrandt.

The code appeared to be partially autonomous; it didn't require anyone to pull the trigger. Instead, it relied on four sophisticated "zero-day" exploits, which allowed the code to spread without human help, autonomously looking for its target.* This fact provided a crucial clue to Chien and O'Murchu: such vulnerabilities are rare commodities, hoarded by hackers, and sold for hundreds of thousands of dollars on the black market. It became clear that Stuxnet couldn't be the work of an individual hacker, or even a team of hobbyists. Only a nation-state could have the resources—and the engineering time—to assemble such a sophisticated piece of code. "It blows everything else out of the water," O'Murchu told me later.

Unsuprisingly, the two men grew paranoid about who might be watching them as they watched the code. Half joking, Chien told O'Murchu one day, "Look, I am not suicidal. If I show up dead on Monday, you know, it wasn't me."

Stuxnet's inner workings harbored another clue that Iran's nuclear program was the malware's target. The worm seemed to be probing for something, in this case a specific kind of hardware known as a "programmable logic controller" made by Siemens, the German industrial giant. These are specialty computers that control water pumps, air-conditioning systems, and much of what happens in a car. They turn valves on and off, control the speed of machines, and watch over an array of sophisticated, modern-day production operations: In chemical plants, they control the mix. In water plants, they control fluorination and flow. In power grids, they control electricity. And in nuclear enrichment plants,

---

* A zero-day flaw is a previously unidentified software vulnerability—so named because there are zero days of notice to get it fixed before the damage is done.

they control the operation of the giant centrifuges that spin at supersonic speeds.

Chien and O'Murchu began publishing their findings in the hope that someone out there was expert in the kind of systems this strange code seemed to be targeting. Their plan worked. One expert in Holland explained to them that part of the code they had published was searching for "frequency converters," devices used to change an electric current, or sometimes change the voltage.

There aren't many innocent explanations for sneaking into someone's infrastructure to change the flow of an electric current. And in Iran's nuclear facility at Natanz, frequency converters played a critical role: they were part of the control system for nuclear centrifuges. And the centrifuges, the US government's experts knew from their own bitter experience, were highly sensitive. Because they spun at supersonic speeds, any dramatic change—triggered, say, by a change in current—could send the rotors out of kilter, like a child's wobbling top. When they became unstable, the centrifuges would blow up, taking out any machinery or people nearby. Uranium gas would be spilled all over the centrifuge hall.

In short, to stop the Bomb, America's new cyber army had made a bomb—a digital one.

As Iran's centrifuges were spinning out of control, the operators at Natanz had no idea what was happening. The data that showed up on their screens seemed normal—the speed, the gas pressure. They had no way of knowing that the code was faking them out and suppressing the signs of imminent disaster. By the time the operators figured out something was dangerously wrong, they could not shut down the system. The malware had affected that process too.

There were other clues. Although the malware eventually infected computers around the world, it kicked into gear only when it found a very specific combination of devices: clusters of 164 machines. That number sounded pretty random to malware sleuths, but it set off my mental alarms. The centrifuges at the Natanz nuclear facility—I knew from years of covering Iran's nuclear program and interviewing inspectors from the International Atomic Energy Agency—were organized in groups of 164.

That left little mystery about the intended target.

The following summer and fall, two *Times* colleagues, Bill Broad and John Markoff, and I published several stories about the hints emerging from the Stuxnet code. Markoff uncovered stylistic and substantive evidence of Israel's role in the code writing. Next, we found one of several American calling cards embedded in the code—an expiration date, when the code would drop dead. Teenagers don't put expiration dates into their code. Lawyers do—for fear that malware could become the digital equivalent of an abandoned land mine in Cambodia, waiting for someone to step on it two decades after it was planted. Finally, Bill Broad discovered the final clue we needed: evidence that the Israelis had built a giant replica of the Natanz enrichment site at their own nuclear weapons site, Dimona. (We didn't yet know the United States was doing the same thing in Tennessee.) The purpose was clear: both countries were building models to practice their attacks, much as the United States built a model of Osama bin Laden's house in Abbottabad, Pakistan, around the same time, to practice the impending raid against the world's most wanted terrorist.

In mid-January of 2011, we felt we had enough information to publish our first story about who had been behind the Stuxnet attacks. In a Sunday article, we laid out the compelling evidence that the United States and Israel had produced the malware together in order to slow Iran's nuclear progress. The story was full of details and markers that took the code right to the gate of Fort Meade, where the NSA is located, but upon publication there was no political outcry, no investigation. That would come more than a year later.[*]

But even after we published our account it was clear there were

---

[*]   The reason for the delay may lie in a coincidence of timing. That first big story was published just hours before Egypt erupted into the chaos of the Tahrir Square uprising, which then occupied all the headlines, and forced President Obama into a tense effort to get President Hosni Mubarak to leave office.

major questions left unanswered: Had this been a small operation gone awry, or a large, well-hidden effort? Assuming the United States and Israel had combined forces to design this enormously complex cyber-weapon, who had given the go-ahead? After all, we knew that in the United States only the president could authorize offensive cyber action, just as he had to provide the launch codes for nuclear weapons.

And if Olympic Games was a sign of where American covert action was headed, were we ready as a nation to open this Pandora's box? Once opened, could it ever be closed again?

THE DISCOVERY THAT Israel had built a replica of the Natanz plant drove home how central a role the Israelis had played in developing the Stuxnet malware. The more sources I interviewed, the more it became clear that the cyber program widened a divide between Prime Minister Benjamin Netanyahu and his brilliant, short, bald spymaster, Meir Dagan. In Dagan's younger days in the Israeli military, he had led squads that hunted down Palestinian militants. Ariel Sharon, the Israeli prime minister who had been Dagan's commander and mentor, famously if crudely declared that "Dagan's specialty is separating an Arab from his head." It was a brutal description, even in the macho world of the Mossad, Israel's best-known intelligence agency, which Dagan ultimately led for nine years—an extraordinarily long tenure. While Dagan pretended to dismiss the stories as mythmaking, he nonetheless seemed to revel in them.

But the mythmaking ignored the fact that it wasn't only Arabs whom Dagan had in his sights. Many observers suspected Dagan's hand in the killing of Iranian nuclear scientists, who were assassinated while driving to work in Tehran traffic after motorcyclists pulled up and attached "sticky bombs" to their car doors before speeding off. If Dagan were indeed behind the killings, it would be in keeping with his view that an Iran armed with a nuclear weapon was truly an existential threat to Israel. Indeed, to talk to Dagan for five minutes was to discover a man who viewed the world through the lens of the Holocaust. On his desk, he kept a photograph of his grandfather kneeling on the ground before his Nazi captors moments before he was killed. It was Dagan's personal "never again" memento that seemed to explain the determination with which he organized the elimination of Israel's enemies.

Dagan made no secret that he never hesitated to send Mossad agents out to kill. Yet when one of those missions went bad—his agents were caught on tape entering and leaving a hotel in Dubai just before and after the 2010 killing of a senior official of Hamas, the Islamist Palestinian group—it was the beginning of the end of his career. The images of the Israeli agents, dressed casually in tennis gear as they entered and left the hotel, played again and again on television. But as his time as the chief of the Mossad dwindled down, Dagan wanted to be remembered instead for managing an operation that was, in his mind, a complete success: the malware attack that crippled Natanz.

Despite Dagan's public reputation as a brutal spymaster who had killed many Arabs in his younger days and ordered the deaths of many more from Mossad headquarters, he was far more strategically savvy than most Israelis knew. Internally, he was increasingly vocal that bombing Iran was madness—it would simply drive the nuclear program further underground. That program would then come back, bigger and more advanced than before. Dagan devoted his last years in office to dissuading Prime Minister Netanyahu from an air attack. "The use of [military] violence would have intolerable consequences," Dagan later told Israeli investigative journalist Ronen Bergman. "If Israel were to attack, Khamenei would thank Allah," he said, referring to the Iranian Supreme Leader. "It would unite the Iranian people behind the project and enable Khamenei to say that he must get himself an atom bomb to defend Iran against Israeli aggression."

All of which meant that by 2010 Dagan was under tremendous pressure to show Netanyahu that a more covert, more sophisticated approach to crippling the Iranian program could succeed.

Dagan and I never met when he was in office. But I was determined to change that after I heard that at one of his retirement parties many of the toasts and jokes made oblique references to the cyberattacks on Natanz. The insiders got the drift; others were left to wonder what everyone else was laughing about.

We first talked in 2011 a few months after Dagan had been ousted from his job by Prime Minister Netanyahu. It was clear he was still bitter about his ouster. He variously derided Netanyahu as a terrible manager and an incompetent warrior. Rightly or wrongly, Dagan believed that Netanyahu had gotten rid of him because the Mossad chief, like other Israeli intelligence leaders, opposed efforts by the prime minister to bomb Iran's nuclear facilities.

"Bombing would be the stupidest thing we could do," Dagan told me. This was not like striking Iraq's Osirak nuclear reactor in 1981 or Syria's reactor in 2007. He believed Iran's program was simply too sprawling; they were not about to repeat their neighbors' mistakes. So while an air attack on Iran's facilities "might make me feel good," Dagan said to me one afternoon, it would provide an illusory solution. The satellite photographs, he said, would show Iran's facilities flattened, and everyone would cheer. But within months, he predicted, those facilities would be rebuilt so deep as to be impermeable to a second strike. And that, he thought, would be disastrous for the state of Israel.

It was fine to try to slow Iran's progress, said Dagan. But if Israel attempted to destroy the country's nuclear facilities in an overt attack, it would ensure a nuclear Iran. There had to be a better way.

A cyberweapon, in his view, was the way out of the conundrum. In our early meetings, Dagan was coy about his role in the development of Stuxnet, even when I mentioned that I had heard about his participation in secure videoconference debates about next steps in the attacks. He didn't know much about computers, he often replied to me with a smile, as if that exonerated him from the role we both knew he played.

Yet over time, as he grew sicker from a failing liver transplant, he edged closer to describing what happened, and why. In our handful of conversations over the years, he sprinkled phrases like "if we did it" into many sentences so that he could explain his underlying logic without violating his oath to maintain the secrecy around the Mossad's covert operations. He talked about how Israel's technology made it enormously difficult for the Iranians to figure out the origin of the attack. The operation against Iran was a model of how Israel should defend itself in the future, he said. Gone were the days of open demonstrations of military might that invited retaliation, escalation, and international condemnation. Gone were the days of occupying territory. The defense of Israel, he insisted, required subtlety and indirection.

"I doubt he knew the first thing about how you write a string of code," one American who dealt with Dagan often told me. "But he knew a lot about how you play with an enemy's mind." And he was convinced that it was the intelligence services that would end Iran's nuclear program, not the air force. It was a mind-set that put Dagan and Cartwright in the same place. And many of Dagan's fellow intelligence chiefs, once they left office, claimed they backed his arguments.

I never heard Dagan directly admit his role in the cyberattacks. But he hinted they had been designed as much to divert Netanyahu from stumbling into a Mideast war as to stop the Iranians from enriching uranium. "I don't trust him," Dagan said of Netanyahu. It was in Netanyahu's interest, he told me, to portray the Iranians as irrational zealots who would use their bomb against Israel. But Dagan looked at the Iranians and saw a group of mullahs mostly interested in staying in power, rather than starting a suicidal war.

There was a reason, Dagan told me, that Bush would not give the biggest bunker-busting bombs to Netanyahu. "He was afraid Bibi would use them," Dagan told me. "And so was I."

That fear explained Dagan's enthusiasm for using cyberweapons against Tehran. It was a way to set the nuclear program back. Perhaps more important, Bush and Obama were able to argue to Netanyahu that there was no reason to bomb while the cyberattacks were working.

The last time I saw Dagan, he chewed me out for what I had written about Olympic Games. But unlike his American counterparts, he complained that I had written too little, not too much.

"You missed a major part of the story," he said, arguing that the Americans had received far too much credit, and the Israelis—and by extension Dagan himself—had received not nearly enough. I had been seduced by Americans who were intoxicated with advertising their own success, he insisted one evening, rather than giving credit to an ally—he carefully didn't say which one—that had done the heavy lifting, gotten the code into the centrifuges, and revised it as necessary.

I would be happy to tell more of Israel's side of the story, I told him, but he'd have to be more explicit about the operational details to prove his point. He smiled—a smile of disgust.

"I am an old man," he said, "and I am sick. I don't want to spend my last days in jail."

By the end of 2011, after dozens of interviews, I had pieced together the highlights of the story of the strategy and debates swirling around the decision to unleash Stuxnet—or at least as many of them as I could gather, given the layers upon layers of secrecy involved. After consulting editors, and the *New York Times* in-house counsel, it was time to go to the Obama White House to see if they were ready to talk—both about what had happened and about any national security concerns they might have about publishing the details. As in all such cases, I made clear that the *Times,* and the *Times* alone, would decide what to publish. But if there were risks to ongoing operations or lives, we needed to discuss them now, not after publication.

My first visit was to Benjamin Rhodes, the former novelist and graceful speechwriter who eventually handled a portfolio of diplomatic issues for Obama, including the opening of Cuba. It was his job to deal with reporters who came to the administration with complex, sensitive national security stories and to decide how, if at all, the White House would respond. Without getting into the details of what happened, he suggested that I go see General Cartwright. It made sense, since Cartwright's term in office spanned the Bush and Obama administrations, and he had been at the center of all offensive cyber debates and understood their sensitivities. Cartwright had retired from the marines after he was passed over for chairman of the Joint Chiefs in 2011 and knew the history of the US military's development of a cyber arsenal better than anyone else.

I knew Cartwright from his days on the Joint Chiefs and had attended conferences where he had discussed the strategic challenges of the new age of cyber conflict. In my reporting on Olympic Games, his name came up often as the man who had tutored Obama about how the Stuxnet worm worked (though it had not yet been given the name "Stuxnet") and rolled out the "horse blanket" diagram of Natanz to bring Obama up to date.

But Cartwright's direct line to Obama had grated on Robert Gates, the secretary of defense, and Mike Mullen, the chairman of the Joint Chiefs. On a variety of issues they believed he manipulated the Pentagon system, or went around the chain of command. It didn't help that Cartwright had not spent time in Iraq and Afghanistan. When Mullen was ready to retire, the two successfully argued against promoting Cartwright into Mullen's role. Suddenly, the man who was among the first to sketch out how the United States could create a dedicated military command to deal with a new dimension of warfare was cast out. With him, I discovered later, went some of the most creative strategic thinking about the use of cyber in offense and defense.

Since retiring, Cartwright had taken a chair at the Center for Strategic and International Studies and signed on with a handful of defense firms, including Raytheon, the maker of missile defenses and defense electronics. Carefully, he had begun to speak out against the secrecy surrounding America's new cyber arsenal, arguing that if the United States wanted to create cyber deterrence it was going to have to show a bit of its capability. "You can't have something that's a secret be a deterrent," I heard him say in more than a few public forums. "Because if you don't know it's there, it doesn't scare you."

He was right, and with the utmost care, the Pentagon began to slip a few lines into public testimony acknowledging that it possessed offensive cyber capabilities. It was a little like acknowledging that the sun rose in the morning. Still, it wasn't an enormously welcome message in the intelligence world, which feared a slippery slope toward divulging what those weapons were and how they were used. Meanwhile, Cartwright was also making a case that the United States could survive quite nicely, thank you, with far fewer nuclear weapons—an argument that carried extra weight coming from the former head of Strategic Command. Again, he was right. But his argument didn't exactly win him friends among his old Pentagon colleagues, who rarely met a weapons system they didn't like.

I took Rhodes's advice and called Cartwright. By the time I went to see him, I not only knew the outlines of the story about Olympic Games, I had already drafted them into two chapters of a book about Obama's first term, describing Olympic Games in as much detail as we could dig up at the time. The book was due to be released in months, and the manuscript was already being edited. That fact became significant later on, when the FBI—whose special agents must have never encountered a book-production schedule before—wrongly concluded that Cartwright was the source of the tale.

My goal in seeing Cartwright was twofold: to check that I had the history and implications right, and to get an independent view of whether any details I was reporting could jeopardize American national security. Cartwright knew that I had been sent by the White House, and saw himself as part of the effort to dissuade me from publishing any details of the operation that might aid an American adversary. He made it clear that he could not discuss classified details. Yet as it turned out when the FBI went looking for a "leaker"—as if there were a single one—we were both a little naïve. For doing what he thought was right, he later paid an awful price, for which I feel enormously guilty to this day.

．　．　．

It WAS JUST a few days later that I made my trek out to the CIA head-quarters to visit Michael Morell in his seventh-floor office—surrounded by three decades of memorabilia from his career, including artifacts from the raid that killed Osama bin Laden. Morell was close to President Obama, and I knew that if the administration was going to push back against the story, he would be the man to do it.

We began pacing through my reporting and the story I was preparing to tell. I ticked through the forensics that led experts to identify the United States and Israel, the carcasses of blown-apart centrifuges found by international inspectors, the mock-ups of Natanz built by the Israelis at Dimona and by the United States in Tennessee. I described the Situation Room debates in which Morell participated. He cautioned against a few assertions and argued with a few conclusions. At a few moments in the story he slowed down, taking notes and suggesting that he might ask that I remove references to certain techniques that the agency used to get malware into target computers and networks. (Curiously, a few weeks later, he asked that a reference to one of those techniques be restored. Though he offered no explanation, clearly the agency had moved on to other methods and wanted to keep the Iranians thinking that the old techniques were still in use.)

In the end, Morell asked for only a handful of deletions, mostly technical details that focused on how the United States put "beacons" and malware into foreign systems and networks. None was essential to telling the story of the most sophisticated state-sponsored cyberattack in history. "You agreed to just about everything we asked for," he acknowledged later on, even while still objecting to the fact that we were publishing anything at all about an American covert operation.

But none of that mattered when the story was published. Republicans who were trying to cast Obama as weak on terrorism—not easy after the killing of bin Laden—accused the White House of leaking the story, along with an unrelated story in the *Times* about the president's role in approving a "kill list" of terrorists to be targeted by drones.

"We know the leaks have to come from the administration. And so we're at the point where perhaps we need an investigation," said Sen. John McCain. He called the story part of "a pattern in order to hype the national security credentials of the president and every administration does it. But I think this administration has taken it to a new level."

Obama himself performed a delicate dance: He couldn't confirm the story, of course, or deny it, but he wanted the world to know he wasn't the source. "I'm not going to comment on the details of what are supposed to be classified items," he said with a hardness in his voice a few days after the details about the White House origins of Olympic Games were published. "When this information, or reports, whether true or false, surface on the front page of newspapers, that makes the job of folks on the front lines tougher and it makes my job tougher— which is why, since I've been in office, my attitude has been zero tolerance for these kinds of leaks and speculation.

"Now, we have mechanisms in place where if we can root out folks who have leaked, they will suffer consequences. In some cases, it's criminal." He quickly added: "The notion that my White House would purposely release classified national security information is offensive. It's wrong."

His comments, made in June 2012, underscored the reflexive secrecy surrounding all things cyber, particularly odd in this case because the code had been spreading around the globe for two years. They also essentially forced the Justice Department to launch a leak investigation, which Attorney General Eric Holder announced around the same time. The White House chief of staff ordered all employees to preserve any notes or emails or communications with me. Since I had been covering the Obama national security team for more than three years, there were a lot of those. Soon the FBI began interviewing scores of potential witnesses. They obtained a secret warrant to get all the emails sent and received by General Hayden, the former CIA and NSA chief. And they used the CIA's notes from my conversation with Morell to try to point the finger at General Cartwright. Why they picked him, out of the scores of officials in the United States and

abroad whom I interviewed, remains a mystery to me. (At one point
they came to him with highlighted lines he had used in speeches, and
the syntax of paragraphs I had written, looking for commonalities. Of
course, all quotations were from Cartwright's public, on-the-record,
unclassified statements.)

As Cartwright himself has since acknowledged, he made an error
of judgment in agreeing to be interviewed by the FBI without a lawyer
present; he said he thought they were all on the same side. When the
interview with the FBI became confrontational, the complaint filed
in his case reported, he became ill and was briefly hospitalized. Later,
when he was indicted, it was for lying to the FBI about when and how
we had met.

He was never charged with leaking any classified information. And
as far as I can tell, he never did. But that crucial fact almost didn't seem
to matter.[*]

The supreme irony of the Cartwright case is that the man who'd
helped propel the federal government into shaping a sophisticated approach to dealing with the world's most complex weapon was among
the first victims of the paranoia about discussing that approach. The
government could have responded to the disclosures about Olympic
Games by embracing the revelations and reminding adversaries—Iran,
Russia, and North Korea among them—that the United States could
do far worse to them. It could have explained why cyber was critical to avoiding a shooting war in the Middle East. It could have used
the moment to talk about what kind of global rules we should create
for using cyberweapons against civilians, against commercial facilities,
and against other governments.

The government did none of that. The Pentagon and the intelligence agencies were unwilling to discuss publicly how they might limit
the use of cyberweapons, in times of both war and peace.

Partly that reluctance reflected the fact that the United States still

---

[*]   In 2016, Cartwright pled guilty. Obama gave him a full pardon in the last
days of his presidency, even restoring his security clearances.

believed it had a lead, if a narrowing one, in cyber technology. In the early days of the nuclear age, many officials had opposed even a discussion of arms control, arguing that there was no reason for the United States to shorten a long lead over its competitors. (The first limits on nuclear weapons happened in the early 1960s, only after the Soviets had a full arsenal, and Britain, France, and China were building them.) But the silence and obsession with secrecy may have had a deeper motivation: American intelligence services had a menu of other cyber operations brewing around the world. These ranged from classic espionage to highly destructive malware—the kind that could knock a whole country back into the analog age.

CHAPTER II

PANDORA'S INBOX

The science-fiction cyberwar scenario is here. That's Nitro Zeus.
But my concern, the reason I'm talking, is when you shut down
a country's power grid, it doesn't just pop back up. It's more like
Humpty-Dumpty. And if all the king's men can't turn the lights
back on, or filter the water for weeks, then lots of people die. And
something we can do to others, they can do to us too. Is that
something that we should keep quiet? Or should we talk about it?
—*An NSA employee, speaking through a composite
character in* Zero Days

AFTER THE RUSSIAN hack of the Pentagon's secret networks in
2008, two things seemed clear to the newly inaugurated Obama
administration. First, Putin's hackers were sure to come back. And second, America needed a full-fledged Cyber Command, far more capable
than the small units spread among the army, the navy, the air force, and
Cartwright's Strategic Command. It was time for a true military organization, with its own troops, that integrated digital offense and defense.

But no one was quite sure what that digital army was supposed to
look like, or how it would wage war. Politicians instantly grasped all
the other battle "domains": land, sea, air, space. They could picture
conventional equipment like tanks, aircraft carriers, bombers, and satellites. But cyber, as Keith Alexander, then the head of the National

Security Agency and ultimately the first commander of Cyber Command, said, "left many of them a bit mystified."

"My grandchildren got it," he told me. "Congress took a little longer."

In fact, Alexander and others found themselves talking to some members of Congress who barely used computers—so it was not easy to explain how a new military force could design malware to defeat an enemy. And while Operation Olympic Games would have provided a vivid example, it was still a highly guarded secret. The operation was so compartmentalized that only a handful of key members had been briefed about its existence.

In 2009, Robert Gates, by then Obama's secretary of defense, concluded after the Russian breach of the Pentagon's classified networks that the creation of US Cyber Command was overdue. It formally came into existence in June, and was housed at Fort Meade—a recognition that, if it wanted to survive, this new military unit would desperately need the skills and experience of the civilian talent at the National Security Agency. Over time a plan emerged to create a 6,200-strong military force—soldiers, sailors, marines, and fliers divided into 133 "Cyber Mission Forces"—that would be spread among the services. A few offensive cyber teams were already housed at Fort Meade, quite explicitly modeled on the Special Forces Command, the favorite of every American president. But turning them into a digital fighting force would take time.

"Special Operations people are hard to find and hard to grow," Ashton Carter told me in 2013, just before he left his post as deputy secretary of defense. But the hardest part, he added, was figuring out exactly what these new forces would be allowed to do. Every US military operation requires the sign-off of lawyers, but figuring out what was permissible under the laws of war was particularly difficult in cyberspace. (This was a uniquely American problem, one that did not slow down the Russians, the Chinese, or the North Koreans.)

"It's things like: Are you sure that a particular action you take with an enemy's information system will only have the consequence of disrupting, let us say, an air-defense system," Carter asked, without shutting down hospitals or cutting off water to civilians? "You have to understand what the consequences are of your actions."

For that reason, Carter added, "these are the kinds of [decisions] that are serious enough that they're reserved for the president." It was a key point: Just as only the president could order the launch of a nuclear weapon, the use of a cyberweapon was similarly limited.

The task of sorting through the rules fell to Keith Alexander, who in turn relied heavily on Paul Nakasone, his chief aide-de-camp. While Alexander was always pushing the envelope—arguing for more authority to collect data flowing into the United States, the way he had done for the digital data flowing into Iraq—Nakasone was immersed in thinking about how to organize a cyber army.

"Everyone who watched him operate—grabbing you in the hall to ask what's going on, fluidly working across the Pentagon and Fort Meade—realized he was being groomed to lead in cyber in the future," recalled Christopher Kirchhoff, a Pentagon aide who went on to be one of the partners in the Pentagon's experimental technology development effort in Silicon Valley.

As it turned out, Nakasone was deeply involved in another critical operation—one of Cyber Command's first big classified projects. It was a subset of what the Pentagon called, in its number-obsessed way, "Op Plan 1025." This was the road map for going to war with Iran, either because negotiations over its nuclear program failed or because Iran lashed out, perhaps in response to an Israeli bombing strike.

Cyber Command's piece of the puzzle was to contribute to an operation named Nitro Zeus. It was a plan—using cyber and other methods—to shut down the entire country, preferably without firing a shot. If Olympic Games was the cyber equivalent of a targeted drone strike on Iran, Nitro Zeus was a full-scale attack.

· · ·

PAUL NAKASONE'S FIRST encounter with computing was not exactly an inspirational Silicon Valley tale of discovery and invention.

"It was 1986, and I bought a PCjr," he recalled. Nakasone was a college student at St. John's University, a small gem of a school on a beautiful lake in a remote part of Minnesota, so remote that the ability to connect to the outside world meant everything. That little computer—with its much-derided "Chiclets" keyboard and its basic operating system—"completely fixated me," he said. Decades later he still remembered the odd combination of commands you needed to make it work. "You know, these were the days when you had to hit 'Control' plus '7' just to print something out. No way you could get much done. But I was hooked."

Nakasone was the son of a Japanese-American linguist who had witnessed Pearl Harbor firsthand. During World War II, his father's language skills solved an immediate wartime need for the government, a service that was enough to keep his family out of the internment camps that the Roosevelt administration had mandated for most Japanese-American citizens. Paul, born nearly twenty years after the war ended, was the first generation in his family to go to college in the United States.

As he tapped the keys on that PCjr in 1986, Nakasone had little inkling of how his first, brief exposure to the new world of personal computing would change his life. When he received his army commission that year, no one paid any attention to his interest in computing—and neither did he. He ran through the traditional posts given to an army career officer interested in rising to the top ranks. That meant thinking the way the army has thought for decades about how to prevent—and fight—a land war.

He did his Second Infantry Division training at Fort Carson in Colorado, followed by a posting on the last border of the Cold War: the Demilitarized Zone, where South Korean and North Korean troops stare each other down as if it were still 1952. From his perch thirty miles north of Seoul, it looked as if the North Koreans could barely make a light bulb. The country was dark.

During the 2008 invasion of Iraq, Nakasone finally got a chance to think digitally. He was part of the "Strategic Initiatives Group," which was just beginning to utilize cyber techniques—no one had yet gotten as far as thinking about it as cyberwarfare—against Islamic extremists. There were a few experiments—infecting laptops and taking down communications lines—but nothing that would get a cyber warrior's blood running.

"The change came in 2008," he told me, when Gates was pushing for the creation of Cyber Command. Nakasone's experience made him a natural to help organize the force. He seemed fluent in a language that left most of his army colleagues a bit dizzy, and more than a little suspicious about the new Pentagon catchphrase, the "digital domain."

Like his father, Nakasone found himself constantly translating for the military—from the code-speak of programmers to the lingo of war planners. "There was the realization, between the Secretary of Defense and the Joint Chiefs, that we needed to think differently about this—to think of it as an entirely new realm of battle," Nakasone told me. Nakasone spent a lot of time explaining that cyber didn't supplant the normal weapons of war. Cyber conflict wasn't separate from every other form of conflict. It would be a part of every future war, and subwar; it would be used right alongside the military's drones and Tomahawk missiles, its F-16s and Special Forces.

But at the beginning, "we didn't have anything," he said. "No structure. No real mission yet. That had to change."

FOR THE NEW troops at Cyber Command, Olympic Games provided a case study in what can go right—and what can go off the rails—when the United States turns to cyberweapons.

The physical damage done by the Stuxnet worm was devastating and dramatic but not long-lasting. By most accounts, the Iranians lost about a thousand centrifuges, and out of fear of further destruction, the Iranian engineers took more offline. But after the code leaked out, they put the pieces together. It took a year of recovery and rebuilding, but they got their capacity back and ultimately installed about eighteen thousand centrifuges—more than three times the number that were installed at the time of the attack. As Iran's foreign minister, Mohammad Javad Zarif, said to me one day in Vienna during the negotiations on the Iran deal, "In the end, what did your vaunted engineers accomplish? They made us more determined than ever to build, and build more."

The attack's more lasting effects were psychological, not physical. When you looked at a chart of Iran's production of enriched uranium, Olympic Games was a blip, not a game changer; a tactical victory, not a strategic one. But it created fear inside the Iranian nuclear establishment.

"The first thing it told the Iranians is that we were way, way inside their systems," one former Israeli official noted later. "That had to make them paranoid. Not only were we inside but we could keep coming back, anytime we wanted. In other words, they could not lock the door.

"The second effect," the official continued, "was that we sent a message. If countries like the United States and Israel were willing to go to these lengths to stop the centrifuges, what lengths would they be willing to go to stop a bomb from being produced?"

And the third message, he said, was that the nuclear program "might be more valuable to them as a bargaining chip than as a bomb-making system."

But as the Iranians rebuilt their program larger than before, President Obama could not count on those messages' convincing the mullahs that it was time to go to the negotiating table to see what they could get in return for giving up their nuclear program. There was always the chance that his effort to restrain Israel would fail, and Prime Minister Benjamin Netanyahu would decide to bomb Iran's facilities—possibly sucking the United States into another war in the Middle East. Obama needed a broader strategy, one that gave him a workable military option.

So even while Olympic Games was under way, Obama ordered up a war plan. In part that decision was driven by Gates, who made clear he was distinctly unimpressed by the quality of the administration's thinking about what the United States would do if Iran raced for the bomb. Gates wrote a blistering memorandum to national security adviser Tom Donilon describing how woefully unprepared the United States was for "strategic surprise."

It fell to Gen. John R. Allen, then at US Central Command—which is based in Florida and oversees the totality of US military strategy in the Middle East—to rectify that deficiency. To this day, General Allen, who went on to lead the Brookings Institution, has never spoken of his efforts there, but the end result was a comprehensive strategy to respond to a nuclear Iran. And Nakasone and Cyber Command had their own piece of that project: integrating cyberattacks with more traditional military operations.

When Nakasone and Cyber Command looked at what their digital weapons could contribute to the battle plan, they focused on the Iranian targets that they could reach by boring into the country's networks: Iran's air defense, its communications systems, and its power grid. Nitro Zeus would be the opening act of the war plan: turning off an entire country so fast that retaliation would have been extremely difficult. It was also, in the minds of some of its creators, a glimpse of the future. The idea was to plunge the target country into blackness and confusion from the very beginning of a conflict. That would give Israel and the United States time to bomb the many suspected nuclear sites, take pictures of how much damage was done, and if necessary bomb them again. But the hope was that Nitro Zeus would avert an all-out war, because the Iranians would, in theory, not be able to strike back. As part of the plan, Iran's missile capability would also have been targeted—an operation whose fundamental concept would return, with a vengeance, as the North Korea crisis heated up.

So, even as President Obama was worried about the vulnerability of America's electric grid, the United States was tunneling inside Iran's grid—along with its cell-phone network and even the Iranian Revolutionary Guard Corps' command-and-control systems.

"This was pretty mind blowing to me," one former official said. "Here we were, going to work every day behind sealed doors, essentially trying to figure out if it was possible to cripple an entire nation's infrastructure without ever firing a shot or dropping a bomb. So we littered Iran's networks with malware," he said, a reference to the process of placing implants in key strategic systems that could, later on, be used to inject destructive code or simply turn the networks off.

"The hard part was keeping track of all of it," he said.

Keeping track was tricky business because networks always change—and because there was no way to test Iran's vulnerabilities in field conditions. So Nakasone and the thousands of people at work on Nitro Zeus resorted to tabletop exercises, simulations of an attack. They tested and retested on a virtual model of Iran's networks to make sure that the implants were not visible to the Iranians and that collateral damage was limited.

And they created answers from scratch to a series of questions: How do you take down the grid and keep it down? How about the air defenses? If the Iranians try to retaliate, how do you make sure they never get off the ground?

"This was an enormous, and enormously complex, program," the former official said. "Before it was developed, the US had never assembled a combined cyber and kinetic attack plan on this scale."

For the United States' cyber warriors, Nitro Zeus was a turning point. It exposed many of the tensions between the National Security Agency—which possessed most of the talent needed to pull off the attack—and the military's newly created US Cyber Command. On paper, the two organizations were complementary. In reality, they had a constant series of spats, typical of arranged marriages, in which the NSA's talent looked down on Cyber Command, and the military unit regarded the NSA as a bunch of arrogant civilians who never needed to complete a military mission.

It was a conflict that would play out time and again. The NSA invested huge resources into getting inside foreign systems, hiding its malware in hard-to-find corners, and checking in on it regularly. Cyber Command usually wanted to grab those implants to conduct attacks—thus revealing their location. "This was the endless squabble," one former member of the NSA said. "It was the difference between intelligence officers, who are in this for the long term, and military officers, who are paid to plan for attacks."

But the most fascinating element of Nitro Zeus might have been not its technical complexity but rather its geopolitical implications. Olympic Games was an intelligence agency–led operation designed to help force Iran to the negotiating table; Nitro Zeus was a military plan, intended to unplug Tehran if diplomacy failed. They both involved cyberweapons but for very different strategic goals.

Taken together, though, the two secret cyber programs suggest how seriously the Obama administration was contemplating the cost of diplomatic failure, and the very real possibility that the US could have found itself in an open conflict with Iran. In the minds of the war planners, that outbreak could have been triggered by something completely out of American control—particularly a decision by Netanyahu to strike Iran's nuclear facilities. "There were many moments when I thought Bibi was on the brink of doing exactly that," Ehud Barak, the former Israeli defense minister and prime minister, told me years later. "And the only question in our mind was, 'If we do it, is the US behind us?'" Nitro Zeus gave America an opportunity to stick with an ally, if necessary, but without committing ground troops, a payoff that had become the holy grail of American power in recent years.

"Nothing else compared to this mission," one insider said later to my colleague Javier Botero. "It was just a huge, expensive undertaking, beyond the reach of anyone but a few nation-states." But apart from the potential implications for Tehran, Nitro Zeus demonstrated the degree to which, in a few short years, Nakasone and his colleagues had transformed America's cyber operations from surveillance tools to vital weapons in the country's arsenal.

· · ·

A PLAN AS big and destructive as Nitro Zeus required the United States to contemplate doing things to Iran's infrastructure that—were they done to us—would be considered an act of war. And the preparations had to be conducted in a way that would not be detected by the Iranians, who would look at the implants in their network and conclude, quite reasonably, that whoever put them there was planning a preemptive attack on their country.

When the US mounted such an operation, the Pentagon called it "preparing the battlefield," and described the moves—if spoken about at all—as a prudent step in case war breaks out. But when the same kind of implants were discovered in American systems, the US was outraged—understandably—and assumed the worst.

"We have seen nation-states spending a lot of time and a lot of effort to try to gain access to the power structure within the United States, to other critical infrastructure, and you have to ask yourself why," said Adm. Rogers, the director of the NSA and the head of Cyber Command until the spring of 2018. "It's because in my mind they are doing this with a purpose, doing this as a way to generate options and capabilities for themselves should they decide that they want to potentially do something."

This, of course, is exactly what we were doing to Iran.

That approach worked in part because Iran was a highly unusual target. The country had so much on the line—global oil sales, investment in the country's broken infrastructure, the ambitions of young Iranians who wanted visas stamped in their passports—that the nuclear program suddenly became negotiable when American and Israeli cyberattacks, combined with sanctions, triggered heated debate in Tehran over whether the country would be stronger as an independent nuclear power or as a major player in the global economy.

The Iranians did not know about Nitro Zeus, although after Stuxnet got loose and Olympic Games was exposed they may have suspected something like it was in the works. But what they did know about American cyberattacks prompted, in combination with the decision to halt their nuclear program, an incredibly foreseeable response: Iran started building a cyber army of its own.

Indeed, while Paul Nakasone's team at Cyber Command burned the candle at both ends preparing Nitro Zeus, the Iranians were already preparing to strike back for Stuxnet. In terms of firepower, their volley would pack little punch compared to the US government's comprehensive plans to shut down their country. Yet, even with their limited cyber capability, the Iranians would expose a difficult truth about cyber conflict, one that Obama would grapple with but never know how to counter: The calculus of offense was inextricably wedded with that of defense. And defending the United States—with its sprawling financial systems, stock markets, utilities, and communications networks, all in private hands—was next to impossible.

WHEN I WAS a kid growing up in the suburbs of New York, we all knew the Bowman Avenue Dam in Rye. It looked more like a toy dam than a real one—twenty feet high, with a single gate. Fed by Blind Brook, it was mostly empty, and thus a great place to clamber around after school. It was also the kind of place your parents probably didn't want you hanging around, for fear you would fall and break something.

I don't think I saw or thought about the dam between junior high school and the day that John Carlin, who headed the national security division of the Justice Department, called me in early 2016. He had just unsealed the indictment of a number of Iranians, with apparent ties to the country's intelligence services, for breaking into Bowman Avenue Dam's command-and-control system in 2013, in what the federal government darkly suggested might be an effort to unleash the water behind the dam to flood a section of New York.

"John," I told him, "I doubt there is enough water in that dam to flood a basement." The idea that this dam even had a command-and-control system was a stretch; my recollection was that the sluice was opened and closed by a big, long bar that was mostly rusted shut. While it was later put under computer control, this wasn't exactly the Hoover Dam.

It turned out that the Bowman Avenue Dam was a mistake for the Iranian hackers; they must have had something like Hoover in mind, and missed. Or maybe it was simply a demonstration of their powers. "The most likely conclusion is that it was a warning shot," Sen. Chuck Schumer, the Democrat from New York, said to me the day of the indictment. The message was, "Don't pick on us, because we can pick on you."

Schumer went on to say that the lesson from this case was "not that we should not employ cyberweapons, but that we should be able to protect ourselves."

If Schumer was right about the retaliatory nature of the strike, it was an interesting insight into one predictable result of Olympic Games. The decision by the United States to make use of a cyberweapon gave the mullahs and the Iranian Revolutionary Guard Corps an excuse to do something they desperately wanted to do anyway: find a pretext for attacking the United States and its allies. To save their pride, if nothing else, they needed to prove they could reach deep inside America's infrastructure and the infrastructure of its allies.

In the summer of 2010, the Iranians publicly announced the creation of a cybercorps to counter the growing US Cyber Command. For historians of the Cold War, this development had a familiar ring: we deployed nuclear weapons, and then the Soviets did; we created bureaucratic structures around those weapons, and then they did.

Following this pattern, after Olympic Games was exposed in 2011, Iranian hackers began targeting roughly four dozen American financial institutions—including JPMorgan Chase, Bank of America, Capital One, PNC Bank, and the New York Stock Exchange. These were not especially creative attacks. Mostly, they were what the government called "distributed denial of service" attacks, often referred to as DDoS attacks, which overwhelm their target with coordinated computer requests from thousands of machines around the world. The targeted networks were never designed to take that kind of volume, and they often crashed, knocking them and any operations that relied on them out of service. Banks were paralyzed. Customers were frozen out of online banking. A group that called itself the Izz ad-Din al-Qassam Cyber Fighters conveniently claimed responsibility.

Nothing about the attack was very sophisticated. "It's primitive; it's not top of the line," James Lewis, the expert on nation-state hacking at the Center for Strategic and International Studies, said at the time. "But it's good enough and they are committed."

With their customers outraged, the banks needed to offer some kind of response, but they quickly found themselves caught up in a central conundrum of American cyber conflict. While Washington urged companies to be more transparent about attacks, high-priced lawyers and security experts offered the opposite advice. Admitting you are a target, they said, just encourages more attacks—and opens companies to liability suits. And for financial institutions trying to convince customers to keep their money there, it was plain old embarrassing. (As seen time and again, even the federal government rarely follows its own advice when its institutions suffer major breaches.)

Most of the targeted financial institutions decided it was better to shut up than to admit the existence of the attacks. JPMorgan Chase, which had openly acknowledged previous denial-of-service attacks, determined this one was so large that it was better to say nothing. Their customers were left in the dark.

The banks weren't the only ones twisted into knots by the problem of what to say. In fact, as the Iranian bank attacks unspooled, the Obama administration struggled to respond. They couldn't simply stay quiet in the face of news that someone was attacking the financial system, yet they were hesitant to elevate the problem to one of national security. While still publicly refusing to say who was behind the attacks, administration officials began inviting bank executives to the White House for emergency briefings. Then they struggled to figure out what this attack actually was. Vandalism? An act of war? Something in between?

In the Situation Room, there were basically two groups of people, one official familiar with the debate recalled. "There were those who said this is the equivalent of an Iranian submarine coming off the coast and launching something." That was the position of some members of the Joint Chiefs, and of some in the intelligence community. "And there were those who said no, it isn't—it's the equivalent of a bunch of Iranians driving down the middle of the street playing a lot of loud music and generally being obnoxious. And you don't shoot at kids who are being obnoxious."

One intelligence official who was involved in the administration's debate admitted there was a lot of distance between those two arguments, explaining: "It was neither one. And this is the problem with our analogies from cyber to the physical world, because . . . it was the kind of attack that undermines the confidence in the banking system of the world's largest economy."

"You can't be in the position of letting someone mess with your banking system," the official went on to explain, "even in a minor way, because the next time it won't be so minor. And that's how you head toward financial chaos."

The Iranians may have done the banks a favor. After the attacks, several officials noted, the financial industry spent billions of dollars building the best cyber protections in any corner of the American private sector.

Still, the bank attacks triggered a familiar debate in Washington: If the United States ever had to strike back, how would it do it? It wasn't an easy question, because the attacks weren't coming out of Tehran; rather, they were coming from servers located in other countries. "When Iran hit our banks, we could have shut down their botnet, but the State Department got nervous because the servers weren't actually in Iran," one former official said later. "So until there was a diplomatic solution, Obama let the private sector deal with the problem."

In fact, Obama was concerned that if the United States came to the rescue of the banks, it would give them little incentive to build their own defenses. At the same time, the White House felt it had to hide the evidence that Iranians were behind the attacks. So that central fact was immediately classified. Congressional staff members were shuttled into secure conference rooms before being told that Iran was the certain culprit, but they were cautioned not to reveal this attribution in public. Of course, as one member of Congress said to me, revealing who was responsible would force a discussion of what the administration was going to do about the attacks. And there were plenty of reasons for the administration's inaction.

It was a ridiculous effort; the secret couldn't last for long. The banks needed to know who had hacked them, and private security teams were beginning to identify the culprits. The government's refusal to say anything about who was behind the attacks only made Washington seem clueless when in fact it knew the answer.

WHAT THE HACKERS were inflicting on American financial institutions, however, was child's play compared to the simultaneous attacks they were launching on rivals closer to home. In midsummer 2012, roughly a year into their active cyber campaign against American banks, Iranians struck Saudi Arabia: their greatest adversary, America's gas station, and the country whose king had suggested to the United States that the way to deal with Iran was to "cut off the head of the snake."

Hackers found an easy target in Saudi Aramco, Saudi Arabia's state-owned oil company and one of the world's most valuable companies. That August, during Ramadan when the Iranians knew most of Saudi Aramco's workers would be away, their hackers wreaked havoc, flipping a kill switch that unleashed a simple wiper virus onto 30,000

Aramco computers and 10,000 servers. Screens went black, and files disappeared. On some computers there appeared a partial image of a burning American flag. In their panic, Saudi technicians ripped the cables out of their computer servers and physically unplugged Aramco offices around the world.

Oil production was not affected. But everything surrounding it was, from the purchases of supplies to the coordination of shipping. For a while, Saudi Aramco couldn't connect with the Saudi Ministry of Energy, with oil rigs, or with the giant Kharg Island oil terminal, through which the Saudis ship much of their crude production. There was no corporate email and the phones were dead.

This was a milestone hack; rather than simply using cyberattacks to disrupt service, Iranian hackers had just proved their ability to utilize malware to inflict physical damage. The wiper software, called "Shamoon," became a model for other countries seeking to conduct attacks for the next few years. While the early evidence suggested the Iranians had simply hacked in, American intelligence agencies quickly concluded that an insider at Saudi Aramco had helped—someone with pretty unfettered access to the oil firm's networks. The Saudis ended up scrapping their infected computers. By one count they bought 50,000 hard drives—basically cornering the world supply—to get back running. It took five months to undo the damage.

IN HINDSIGHT THE Iranian counterattacks, from Saudi Arabia to Wall Street to the decrepit Bowman Avenue Dam, were more than just tit-for-tat. They were our first look at what low-level, never-ending cyber conflict looks like.

Like the skirmishes at the DMZ and in East Berlin during the Cold War, these attacks did not seem likely to escalate into a broader war. Instead, everyone hewed to the unspoken rules about keeping the cyber conflict just below the line that could trigger armed conflict. The US went for what the Iranians valued most—their nuclear program—and

Iran went for what America valued most: its financial markets, its access to oil, and its sense of control over its own infrastructure. There was disruption and signal-sending. But no one got killed.

"We spent a lot of time on the Saudi Aramco hack, and the Iranian use of cyber in general," said one senior intelligence officer who had spent a career studying the Middle East. "You have to think of their pyramid of weapons." He formed his thumbs and forefingers to illustrate—showing the three sides of a triangle. "We're used to thinking nuclear on top, then bioweapons, then maybe chemical weapons and just ordinary firearms. But they've put cyber on the top—above all of that."

Why? Not because cyberweapons are as devastating as nuclear or biological weapons—except in the most extreme of cases. "It's because the things you need to run a modern society and oil economy—to run Saudi Arabia—all depend on electricity, valves, and pipelines. And the Saudis are incredibly vulnerable and can't solve this problem." Their distribution networks were built up over decades and are connected to other countries. So are the control systems for those networks. The Iranians don't have to get into the system through Saudi Arabia; there are more entry points around the Middle East than they can count. "And the Iranians have calculated that the Saudis aren't going to go to war for an oil disruption whose origin they can't prove. That's the theory, anyway."

The Saudi Aramco attacks were also an early lesson in some of the conundrums of American vulnerability: While the United States devised, at some expense, a vast plan to shut down Iran in the event a conflict broke out, Nitro Zeus sat on the shelf, unused. No doubt it became a model for integrating cyberweapons into American war plans against other potential enemies. In the meantime, Iran, with far less reach and capability, perpetrated small attacks that exposed not only the ease of creating problems for American businesses but also the vulnerability engendered by American secrecy surrounding cyber.

Because the United States never talked about its own attacks on

Iran, it became virtually impossible to debate publicly the wisdom of the original decision to go after Iran's infrastructure—to ask what Robert Gates called the least-asked question in Washington: "And then what?"

America's secrecy about offensive cyber, and its fear of revealing sources and methods, meant that the government never really warned American banks and businesses that they were ripe targets for the new Iranian cybercorps. Instead, the United States issued general cautions about the need for cyber defenses and information-sharing—the digital equivalent of telling people to seek shelter in their basements in the case of a nuclear exchange without mentioning that it would be the radiation, as much as the blast, that was likely to wipe them out.

"Wasn't this ridiculous?" I asked one of Obama's senior aides during the Iranian attacks. If we had bombed an Iranian airbase, wouldn't we warn Americans about the specific threat of retaliation?

"We didn't want to scare people about something they really couldn't do much about," came the response. The official went on to explain that chief executives had been given special clearances, as if that relieved the government of responsibility. But it turned out there wasn't much those chief executives could do with any privileged information they learned.

"I couldn't tell my own information technology managers what I had heard," one of them said when I asked about the briefings. "There was literally nothing I could do with this information except stay up at night and worry about it."

Of course, when failure of network defense is already an issue, secrets have a way of getting out. Soon, secrets far larger than who was attacking America's banks would be wrenched out of the government's hands. All it took was a single contractor at the NSA with grievances about the government, a substantial ego, and easy, unmonitored access to the agency's deepest secrets.

CHAPTER III

THE HUNDRED-DOLLAR TAKEDOWN

Did the combination of Snowden, Cyber Command, Stuxnet . . .
spark a panic among U.S. adversaries and a subsequent arms
race in offensive cyber operations that is adversely affecting
the United States? . . . Hard questions, scary times.

—*Jack Goldsmith, Harvard Law School professor and former assistant
attorney general, Office of Legal Counsel to President George W. Bush*

T HE FIRST BIG public revelation of the National Security Agency's
deepest secrets, and the most costly blow to its multibillion-dollar
programs to break into computer networks from Tehran to Beijing to
Pyongyang, occurred thanks to a piece of commercial software called
a "web crawler." Retail price: under $100.

Web crawlers are exactly what they sound like. They are essentially
digital Roombas that move systematically through a computer network the way that Roombas bounce from the kitchen to the den to
the bedrooms, vacuuming up whatever lies in their path. Web crawlers
can automatically navigate among websites, following links embedded
in each document. They can be programmed to copy everything they
encounter.

This particular web crawler was deposited in the NSA's networks
in the spring of 2013 by Edward J. Snowden, a Booz Allen Hamilton
contractor working at an NSA outpost in Hawaii. Perhaps the most

astounding aspect of his effort to steal a huge trove of the agency's documents—a move considered treasonous by many, but long-overdue and patriotic civil disobedience by his supporters—is that it worked so well: the world's premier electronic spy agency was completely unprepared to detect such a simple intruder swimming in a sea of top-secret documents.

In its embarrassment, the best excuse the agency could muster was that the process of updating security measures in their more far-flung outposts hadn't yet reached NSA Hawaii—more formally, the Hawaii Cryptologic Center near Wahiawa, on Oahu. "Someplace had to be last" to get the security upgrades, one of the agency's top executives told me somewhat sheepishly.

If you take Snowden at his word, his goal in revealing the inner secrets of the NSA was to expose what he viewed as massive wrongdoing and overreach: secret programs that monitored Americans on US soil, not just foreigners, in the name of tracking down terrorists who were preparing to attack the United States. The vast databases at NSA Hawaii disgorged several examples of programs that bolstered Snowden's case that the NSA had used the secret Foreign Intelligence Surveillance Court, and compliant congressional committees, to take its surveillance powers into domestic phone and computer networks that at first glance seemed off-limits by law.

But the main focus of the NSA division within the secure complex in Hawaii was across the Pacific. From Hawaii, not far from Pearl Harbor and the US Pacific Command, the NSA was deploying its very best cyberweapons against its most sensitive targets, including North Korea's intelligence services and China's People's Liberation Army. The weapons ranged from new surveillance techniques that could leap "air gaps" and penetrate computers not connected to the Internet to computer implants that could detonate in time of war, disabling missiles and blinding satellites. While the American public and much of the media were transfixed by the image of a "Big Brother"—tracking not only the numbers they call but the trail of digital dust left by the smartphones in their pockets—the most revealing documents in Snowden's trove showed the vast ambitions of the nation's new cyber arsenal.

If the revelation of Operation Olympic Games had given the public a peek through the keyhole at America's most sophisticated offensive cyber capabilities, Snowden offered the Google Satellite view, from miles above. From there, it was a remarkable sight. It was immediately clear that, over the past decade, the United States had tasked thousands of engineers and contractors, working under tight secrecy, to build a range of new experimental weapons. Some of these weapons merely pierced foreign networks and offered another window into the deliberations and secret deals of adversaries and allies—basically a cyber-assisted form of traditional espionage. But other tools went much further, by burrowing deep into foreign networks that one day the United States could decide to cripple or destroy. The trove of stolen classified documents contained only hints of these programs, since Snowden's access had only got him so far. But taken together, they strongly suggested that Nitro Zeus had been just the beginning.

AT ONCE NAÏVE and cunning, brilliant, articulate, and highly manipulative, Snowden grew up in a North Carolina family steeped in military tradition. A diffident student, Snowden briefly attempted the Special Forces training program before washing out. By his mid-twenties, he had bounced around several colleges, dabbled in Buddhism, and developed a fascination with Japan. He would later describe his first job as a security guard for the NSA—a modest exaggeration. He did not, in fact, work at Fort Meade. Rather, he had guarded a nearby university research center linked to the NSA.

Snowden's big break came in 2006, at age twenty-three, when the CIA needed quick hires to fulfill their growing counterterrorism mission. They moved him into a telecommunications job, then sent him—undercover, no less—to Geneva. Three years later he quit, rightly concluding that his expertise would be more profitable in the private sector. He took a position at Dell, advising the NSA on updating its computers, and ultimately ended up in its Hawaii operations. But his real ambition was to work inside the NSA.

Either to cover his tracks or because he sincerely believed it, Snowden had professed early in his career that he hated leaks, leakers, and the news organizations that make use of their material. The *Times* was on the wrong side of his wrath in 2009, when I revealed in an article that President Bush, in turning down the Israeli request for bunker-busting bombs to deal with Iran's nuclear program, initiated a secret program to attack the country's computer networks—what later became "Olympic Games."

"HOLY SHIT, WTF NYTIMES," Snowden wrote that day in 2009. "Are they TRYING to start a war? Jesus Christ. They're like WikiLeaks."

The man who would later revel in revealing scores of sensitive programs sounded incensed.

"Who the fuck are the anonymous sources telling them this? Those people should be shot in the balls."

But somewhere along the line Snowden's views about the importance of shining a light on America's hidden battles in cyberspace underwent a radical transformation. "It was seeing a continuing litany of lies from senior officials to Congress—and therefore to the American people . . . that compelled me to act," he posted in another online chat, after he had fled the United States.

Snowden's motives at that time remain a subject of vociferous debate. But he wanted a job at the NSA so badly that he broke into a government computer system to swipe the admissions test. Armed with the answers, he aced the exam. Yet when the offer from the NSA came, Snowden was insulted—it was for a midlevel bureaucratic position, with a midlevel salary to match. So he applied for the next best thing: a job at Booz Allen Hamilton, the company that, behind the scenes, had designed many of the NSA's most important computer systems and provided the staff to keep them running.

Snowden would have a contractor's badge hanging around his neck, not the NSA badge he coveted. But he would have as much overview as an employee—and more, once he got his hands on a higher-level password.

As a result, the files Snowden scooped up with the $100 web crawler included what Rick Ledgett, the NSA's number-two official, described to me as "the keys to the kingdom."

It was Ledgett, an NSA veteran, who pulled the short straw to run what the agency misleadingly called its "Media Leaks Task Force," as if the NSA's Snowden problem had instead been about the newspapers and broadcasters who spread the agency's secret documents around the globe. When I suggested to him once that a more appropriate name might have been the "Insider Threat and Internal Mismanagement Task Force," Ledgett smiled and demurred.

"Snowden was part of the problem, but not the only problem," he argued. But Ledgett acknowledged the central fact: the agency "had no idea" that a web crawler you could buy on Amazon had spent weeks working its way through an estimated 1.7 million documents in the agency's systems.

Exactly how many documents, PowerPoint slides, and databases Snowden copied and smuggled with him when he fled Hawaii for Hong Kong is still a matter of dispute. Most have never been published. But almost all were unencrypted because of an assumption at the time—a remarkable assumption, given the nature of the agency's mission—that if you were inside the NSA system, you were trusted and could copy just about anything without setting off alarms.

Luckily for the NSA, Snowden didn't have all the kingdom's keys. The agency compartmentalizes its data into multiple levels, and Snowden only reached the documents that describe the agency's programs, but not the specific sources or details of the tools that enable them. But that left plenty to reveal: programs with names like PRISM—that allowed court-approved, if limited, access to the online Google and Yahoo! accounts of tens of millions of Americans. And there was XKeyscore, which offered sophisticated new methods to filter vast flows of Internet data that the NSA tapped into as the data moved around the globe.

The Snowden trove was like an archeological dig through the innovations of the past decade. He had unearthed how the NSA was working to break the encryption of cell-phone data, and how it was undermining even the "virtual private networks"—or VPNs for short—that companies and many computer-savvy users had turned to in hopes of protecting their data. Those private networks turned out to be not quite as private as advertised.

As one of my *Times* colleagues put it so well, the agency had become "an electronic omnivore of staggering capabilities."

The documents made clear that we are in a golden age of digital surveillance. The United States was collecting what a presidential commission later called "mass, undigested, non-public personal information" about Americans just in case it wanted to mine that data sometime in the future "for foreign intelligence purposes." Within the United States debate raged over whether the NSA had overstepped its bounds in sucking up vast amounts of data on Americans, virtually washing away any distinction between "domestic" and "foreign" communications.

But from the beginning America's allies and adversaries learned something far more important from the Snowden documents—namely, that the NSA's interests in global surveillance and sabotage went well beyond Iranian centrifuge plants. Snowden gave these observers two gifts. The first was an understanding of the NSA's global operations from Berlin to Beijing. The second was an excuse for countries around the world to attempt to stymie American technological dominance in their markets.

The starkest lesson to emerge from the Stuxnet and Snowden experiences is that the cyber world still operated with almost no internationally accepted rules of behavior. It was, as Obama put it, the "wild, wild West," in which countries, terrorists, and tech companies constantly tested the boundaries with few repercussions.

China and Russia would use the Snowden revelations to justify draconian rules that require any American company operating within their borders to turn over pictures, emails, and chats on demand—essentially cooperating in the perpetuation of an authoritarian state.

As for Europe, after the Germans realized the NSA was keeping a busy surveillance office open on the rooftop of the American embassy overlooking the famed Brandenburg Gate, they began talking about creating a "Schengen" routing system in which online data would be kept in Europe. It would be designed to defend against their ally, the United States, more than against their Russian adversaries. The idea was enthusiastically endorsed by Chancellor Angela Merkel, whose cell phone was an NSA target. But the plan was also technologically ill conceived, and eventually the Germans discovered it would do nothing to prevent the NSA from hacking into their networks. In fact, it might even make it easier.

A HALF DECADE after the Snowden revelations, it is remarkable how many questions the NSA was never forced to answer in public. Its officials were able to hide behind the secrecy that surrounds its operations, even though the Snowden trove gave the world an unparalleled look at their work. Publicly, the intelligence agency leadership treated the entire Snowden insider leak as equivalent to a natural disaster: something you regretted but couldn't do anything about.

James Clapper said Snowden had taken advantage of "a perfect storm" of security lapses, yet no one was publicly blamed for those lapses. "He knew exactly what he was doing," Clapper said. "And he was pretty skilled at staying below the radar, so what he was doing wasn't visible."

Modest changes were made after Snowden's revelations. The NSA briefly cracked open its doors, recognizing that it had to explain itself to the American people. One civilian NSA worker was removed from his job—presumably the one who'd let Snowden use his higher-level passwords—and a contractor and a military officer were blocked from access to NSA data. But the agency offered no information on Snowden. No one wanted to explain in too much detail what had happened, or who should be held accountable. The era of "No Such Agency," as its employees used to call it, only half in jest, had to end, but accountability had its limits.

Neither Booz Allen nor the NSA ever explained how Snowden could load vast amounts of data on some kind of electronic device—he has never said what it was—and walk out the door, uninspected and un-impeded. While intelligence officials kept darkly hinting that Snowden must have been working for the Chinese or the Russians all along, they never offered proof that he was a sleeper agent burrowed deeply inside the NSA. Instead, they whispered suggestions to reporters that they examine where Snowden was staying in Hong Kong and look into how his exile in Russia was arranged. But if officials collected evidence that the world's two other superpowers had placed Snowden in the midst of NSA Hawaii, they have never offered it—perhaps because they hope to prosecute him someday, or perhaps because it would be so supremely embarrassing.

Most important, the NSA has never had to account for the fact that it ignored so many warnings about its well-documented vulnerabilities to a new era of insider threats. The warnings had been quite public. Only three years before the Snowden breach, an army private now known as Chelsea Manning had gotten away with essentially the same thing in Iraq—downloading hundreds of thousands of military videos and State Department cables and handing them off to WikiLeaks.

Shortly after the Snowden fiasco, the agency announced new safe-guards: No longer would systems administrators with access to vast databases be able to download documents by themselves. There would now be a "two-man rule"—reminiscent of the dual keepers of the keys for the launch of nuclear weapons—to protect against lone actors.

But the NSA's solution was either too late, or ineffective. Over the next few years, the NSA demonstrated time and again that it could not keep its own secrets. Snowden is simply the most famous insider so far.

. . .

IN THE DAYS after Snowden showed up in Hong Kong and began dol-
ing out parts of his trove of government secrets to the *Guardian,* the
problem of reliance on outside government contractors took center
stage: why was the US government depending on Booz Allen to run its
most sensitive intelligence operations? Sen. Dianne Feinstein, then the
chairwoman of the Senate Intelligence Committee, told me in 2013
that soon "we will certainly have legislation which will limit or prevent
contractors from handling highly classified and technical data." That
never happened.

Astoundingly, Booz Allen never made public why it assigned
Snowden to such a sensitive set of tasks, and why it left him so loosely
supervised that he could download highly classified documents that
had nothing to do with his work as a system administrator. Nor did the
firm lose any of its contracts with the NSA.

Furthermore, the talk in Washington about cutting down on the
use of contractors who dealt with the nation's deepest secrets fizzled almost immediately. "We had to go to Congress and quietly explain how
cyberweapons get developed," one NSA official said to me. In short,
the NSA told Congress, cyberweapons get built the way everything else
gets built—by private firms. The Pentagon relies on Lockheed Martin
to build the F-35, with a raft of subcontractors and partners. General
Atomics builds the Predator and Reaper, the two best-known drones.
Boeing builds satellites. Booz Allen, and many firms from the outskirts
of Fort Meade to Silicon Valley, build cyberweapons.

Those firms hunt for—or surreptitiously purchase—"zero-day"
flaws: software flaws in a system that an invader can exploit to spy
or destroy. The programming talent required to turn those flaws into
weapons is expensive, and contractors can afford top talent. They offer
their best coders salaries many times what the government can pay.
"People would be shocked," one young employee of one of the most
successful cyber contractors told me, "how much the government relies

on contractors to build the weapons and maintain them," even in foreign systems. That explains why about one-third of the 1.4 million people with top-secret clearances in 2012 were private contractors. (And yes, the background checks for those contractors are often performed by other contractors.)

The man who put Booz Allen on the NSA's map was J. Michael McConnell, who knew life on both sides of the revolving door of the cyber-industrial complex. A former navy intelligence officer who made his name in the backwaters of the Mekong Delta during the Vietnam War, McConnell was a pale and stooped man with wire-rim glasses, giving him a vague resemblance to George Smiley, the complex character at the center of John le Carré's novels about life in the bowels of British intelligence. Indeed, McConnell looked more like a bureaucrat than a cyber warrior. Looks were deceiving.

In McConnell's years as director of the NSA under President Clinton, he became increasingly concerned about America's growing cyber vulnerability. By the time he returned to government as director of national intelligence under President Bush, he was ready to enter a new arms race. In Bush's second term, it was McConnell, along with Hayden and Cartwright, who pushed the country into the business of sophisticated cyber projects by overseeing Olympic Games and other offensive operations. And when Barack Obama was preparing to take office, it was McConnell who briefed him on America's covert actions abroad, from Afghanistan to Iran.

After he left his post, McConnell returned to Booz Allen in 2009—for a whopping $4.1 million his first year back—to ramp up its cyber capabilities for the era of conflict he clearly saw coming. He pressed for the development of "predictive" intelligence tools that companies and governments could use to scour the web for anomalies in behavior that could warn of an approaching cyber or terror attack. The work paid off for Booz Allen: Just before the Snowden revelations, the company won a $5.6 billion contract to conduct intelligence analysis for the Defense Intelligence Agency, and another $1 billion from the navy to help with

"a new generation of intelligence, surveillance, and combat operations."
Similarly, when the United Arab Emirates decided to shop around for
a complete signals-intelligence, cyber-defense, and cyberwarfare unit of
its own—one that would befit its role as one of the United States' closest partners in the Arab world—it turned to Booz Allen, and specifically to McConnell, to assemble it. "They are teaching everything," one
senior Arab official explained to me. "Data mining, web surveillance,
all sorts of digital intelligence collection."

McConnell made a persuasive case to me during Bush's time in
office that private firms are needed to jolt the government out of its
attachment to old systems. The air force, as he pointed out frequently,
fought the concept of drones for years.

But he also argued in 2012, just before the Snowden debacle forever
changed Booz Allen's reputation, that the private sector had to take the
security of its own systems more seriously. "It should be a condition for
contracts," he said. "You cannot be competitive in the cyber era if you
don't have a higher level of security."

He may have been wrong there. With the subsequent arrest of another of his employees, the company's stock took a brief plunge on the
expectation that the firm was in trouble. Then the stock bounced back
to all-time highs as investors poured their money into a sure bet to
profit from the newest arms race.

Just as some companies in America proved too big to fail, some
contractors were simply too important to ditch.

For months, Snowden's disclosures roiled Washington. Unlike the
CIA, the NSA had never been plagued with insiders or double agents.
There was both real outrage and faux outrage about the kind of data
the NSA was retaining—but insisted it was rarely looking at—about
American citizens.

The Snowden revelations that got some of the biggest headlines in
the United States revolved around a single document: a copy of the

"Verizon order" from the Foreign Intelligence Surveillance Court. It revealed that the secret court had adopted a legal theory that the USA PATRIOT Act—passed in the days after 9/11—could be interpreted to require Verizon and other big carriers, like AT&T, to turn over the "metadata" for every call made into and out of the United States. And then, it added, for good measure, all calls "wholly within the United States."

"Metadata" do not include what callers said to their husband, boss, or kids. They are a record of the numbers called, how long the call lasted, and how it was routed. Today, nearly two decades after 9/11, collecting that data for all Americans, on their home lines and their cell phones, seems like a pretty clear case of surveillance-state overreach. And it was: the flood of data was so large that it was of extremely limited use. But its collection was a prime example of how the 9/11 attacks so bent the judgment of otherwise smart officials that they began hoarding all kinds of information simply because one day it might be useful—likely not thinking much about the precedent they were setting around the world, especially in countries like China and Russia whose leaders were looking for any excuse to tighten the noose on dissent.

The more troubling problem about the phone-call metadata program was not how it was being used but rather that a series of American officials had lied before Congress about its existence. Snowden's revelations exposed them. The incident was yet another illustration of the way the over-secrecy surrounding how the government uses its cyber powers forced officials to attempt to conceal a program that could easily have been made public without harming its effectiveness.

Yet in the end Snowden's biggest impact wasn't in the defense of privacy. For all the talk on Capitol Hill and cable television about reassessing the balance between security and privacy rights, little changed. Congress renewed the NSA's surveillance powers, with very modest adjustments.

The public debate around privacy issues obscured what was truly revelatory in the Snowden trove: PowerPoint after PowerPoint documented how the NSA's Tailored Access Operations unit—known as

TAO—found ways to break into even the most walled-off, well-secured computer systems around the world.

Informally, everyone still uses the name TAO. Formally, the TAO no longer exists; it has been absorbed into the agency's other offensive units. It began small, but since its founding two decades ago, it has grown into the agency's most storied unit, spreading more than a thousand elite hackers over sites from Maryland to Hawaii and Georgia to Texas.

The unit cannot compete with Silicon Valley salaries, but the mission is irresistible. The hacking unit attracts many of the agency's young stars, who thrill to the idea of conducting network burglaries as a form of patriotic covert action. And their target list is vast: Chinese leadership, Saudi princes, Iranian generals, German chancellors, North Korea's own Reconnaissance General Bureau. Much of TAO's ouput is labeled "exceptionally controlled information," and often makes it into the President's Daily Brief.

There is a pecking order in the TAO, its alumni say. The veterans plot ways to get into foreign networks, then hand the operational challenge off to more junior members of the team who spend days and nights "exfiltrating" information the way the CIA used to exfiltrate foreign spies. "Sometimes it happens quickly; sometimes we take a long time," one former senior member of the organization told me. And what works this week may not work the following week.

That is why TAO workers are constantly designing new malware implants that can lurk in a network for months, maybe years, secretly sending files back to the NSA. But the same implant can alter data or a picture, or become the launching pad for an attack.

And one of the juiciest targets for TAO, it turned out from Snowden's reports, was a country that Americans worried was targeting us: China.

FOR YEARS, AMERICAN officials had considered Huawei (pronounced WAH-way), the Chinese telecommunications giant, a huge security threat to the United States. It feared that Huawei's equipment and products—everything from cell phones to giant switches that run telephone networks to corporate computer systems—were riddled with secret "back doors." Classified intelligence reports and unclassified congressional studies all warned that one day the People's Liberation Army and China's Ministry of State Security would exploit those back doors to get inside American networks.

In 2005 the air force hired the RAND Corporation to examine the threat from Chinese networking firms. Huawei was high on the list of threats: RAND concluded that a "digital triangle" of Chinese firms, the military, and state-run research groups were working together to bore deeply into the networks that keep the United States and its allies running. At the center of the action, they suggested, was the founder of Huawei, Ren Zhengfei, a former PLA engineer who, the Americans suspected, had never really left his last job.

Little proof was offered, at least in public. Nonetheless, the word went out across Washington: Buy Chinese equipment at your peril. There was no hope of banning the company from global networks, as Huawei was on course to becoming the largest telecommunications equipment company in Asia—its inexpensive phones are ubiquitous from Beijing to Mandalay—and the third largest in the world. Its equipment, down to the chips used in cell phones, was integrated into products around the world, from Britain to South Korea. The firm boasted that it connected one-third of the world's population. But egged on by Huawei's American competitors, Washington decided to draw a firewall around the United States. When Huawei tried to buy 3Com, a failing American firm, the Committee on Investments in the United States— a little-known government agency run as an offshoot of the Treasury Department—blocked the purchase on national-security grounds.

In classified briefings to Congress, the NSA laid out its fears: It was almost impossible to know what hidden capabilities Huawei could etch into its hardware, or bury in its software. If there was a hot war with China—or even just a nasty regional dispute—Huawei might be the vehicle for shutting down servers or crippling the US telecommunications grid. Once the telecommunications system was corrupted, other networks would follow. And there was always the fear of theft: What better way to route secret communications to the PLA than through an already established phone company?

The paranoia was not limited to Huawei, of course. After Lenovo, a Chinese computer upstart, bought IBM's personal computer division in 2005, the State Department and the Pentagon largely banned its indestructible laptops. But Huawei, because of the dominance of its products, was a constant focus of investigations by the House Intelligence Committee and American intelligence agencies. The problem was that they could cite no evidence, at least in their unclassified reports, to confirm their suspicions that the Chinese government pulled the company's strings or ordered it to sweep up data. (That did not stop the House from concluding that Huawei and another Chinese company, ZTE, must be blocked from "acquisitions, takeover or mergers" in the United States and "cannot be trusted to be free of foreign state influence.")

The apparent absence of evidence gave birth to "Shotgiant."

That was the name of a covert program, approved by the Bush White House, to bore a way deep into Huawei's hermetically sealed headquarters in Shenzhen, China's industrial heart. And while American officials would not describe it this way, the essential idea was to do to Huawei exactly what Americans feared the Chinese were doing to the United States: crawl through the company's networks, understand its vulnerabilities, and tap the communications of its top executives. But the plan went further: to exploit Huawei's technology so that when the company sold equipment to other countries—including allies like South Korea and adversaries like Venezuela—the NSA could roam through those nations' networks.

"Many of our targets communicate over Huawei-produced products," one NSA document describing Shotgiant reported. "We want to make sure that we know how to exploit these products," it added, to "gain access to networks of interest" around the world.

There was another goal as well: to prove the American accusation that the PLA was secretly running Huawei and that the company was secretly doing the bidding of Chinese intelligence.

The American concern about Huawei was justifiable. No country had made more of an effort to get deep inside US networks than China. "China does more in terms of cyber espionage than all other countries put together," the expert James Lewis noted to me in the midst of the investigation into Shotgiant. "The question is no longer which industries China is hacking into. It's which industries they aren't hacking into."

So Huawei was a natural source of concern. Any firm built in an authoritarian, government-takes-all environment is going to turn over to the state whatever data it is told to turn over. The same worries, officials told me, applied to Kaspersky Lab, the Russian antivirus software maker, whose products were making it easy for Russian intelligence agents to exfiltrate secret American documents.

But when the German weekly *Der Spiegel* and the *Times* published the details of Shotgiant, based on Snowden documents, the depth of the hypocrisy struck not only the Chinese but also many American allies. "You are essentially doing to the Chinese exactly what you are accusing them of doing to you," one European diplomat, whose country was also wrestling with the Huawei problem, said to me one morning over breakfast. He paused for a minute. "Fair enough," he said. "We should probably help you."

Naturally, the American officials who were willing to talk about Shotgiant when it was revealed in 2013 had a different explanation. The United States, they argued, breaks into foreign networks only for "legitimate" national-security purposes. "We do not give intelligence we collect to US companies to enhance their international competitiveness or increase their bottom line," said Caitlin Hayden, then the spokeswoman for the National Security Council. "Many countries cannot say the same."

The problem was that the Chinese did not distinguish between "economic advantage" and "national security advantage." To a country whose power rests on keeping the economy growing, there is no such distinction. Chinese officials looked at the American explanation as self-serving at best, and deceptive at worst. Clearly, one senior Chinese diplomat assigned to Washington at the time argued to me, the NSA's real purpose "is to stop Huawei from selling their equipment so that Cisco can sell its own."

The slides explaining Shotgiant in the Snowden trove gave a sense of the NSA's thinking: "If we can determine the company's plans and intentions," an analyst wrote, "we hope that this will lead us back to the plans and intentions of the PRC." The NSA saw an additional opportunity: As Huawei invested in new technology and laid undersea cables to connect its networking empire, the agency was interested in tunneling into key Chinese customers, including "high priority targets—Iran, Afghanistan, Pakistan, Kenya, Cuba."

In short, eager as the NSA was to figure out whether Huawei was the PLA's puppet, it was more interested in putting its own back doors into Huawei networks. It was a particularly important mission because the Chinese firm was popular in hard-to-access countries where American telecommunications companies were unlikely ever to get a contract. In other words, Huawei might serve as a back door to the PLA, but it would also be host to another back door, one it didn't know about: to the NSA.

By the rules of spy craft, that is entirely in-bounds. America spies, China spies, Russia spies—and turning the tables is fair play. But there is a long-term cost that Snowden, perhaps unintentionally, highlighted. If the United States wanted to set rules for the rest of the world to play by—starting with not exploiting technology that can undermine critical infrastructure—it would have to be willing to give something up. And as Shotgiant made clear, no one in the NSA, or beyond, wanted to contemplate what that might be.

THE CHINA FILES showed that Huawei was hardly the NSA's only target. In 2013 the agency cracked two of China's biggest cell-phone networks and was happy to discover that some of the most strategically

important units of the Chinese Army—including several that maintain its nuclear weapons—were overly dependent on easy-to-track cell
phones. Other Snowden documents laid out how the NSA had mapped
where the Chinese leadership lives and works. There was a huge bull's-
eye on Zhongnanhai—the walled compound next to the Forbidden
City that was once a playground of the emperors and their concubines.
Today it is a mix of ancient splendor and some aged-looking suburban
homes with—at least until the Snowden revelations—ill-protected
Wi-Fi networks. It turned out that the Chinese leadership, like everyone
else, were constantly complaining about how slow their Wi-Fi was and
upgrading their equipment. That created an opportunity for the NSA.

It was one that the Tailored Access Operations unit was prepared
to tackle. In late 2013, *Der Spiegel* published the "ANT catalog," an
equipment catalog that James Bond might have admired.

Starting in 2008 or so, the NSA began making use of new tools
designed to steal or alter data in a computer even if it is not connected
to a network—exactly what it did in Iran to get past the "air gap" that
separated the Natanz plant from the digital world.

The most ingenious of the devices relied on a covert channel of
low-frequency radio waves transmitted from tiny circuit boards and
USB keys inserted surreptitiously into the target computers. Getting
the equipment into the computers required, of course, that the United
States or one of its allies insert the hardware into the devices before
they were shipped from the factory, divert them while they were in
transit, or find a stealthy spy with a way to gain access to them—no
easy task. But sometimes it was also possible to fool a target into inserting the devices themselves. The ANT catalog included one device,
called Cottonmouth I, that looked like a normal USB plug that you
might buy at Office Depot. But it had a tiny transceiver buried inside
that leapt onto a covert radio channel to allow "data infiltration and
exfiltration."

Once the illicit circuitry was in place, the catalog indicated, signals from the computer were sent to a briefcase-size relay station—

wonderfully called "Nightstand"—that intelligence agencies could place up to eight miles away from the target. In other words, an American intelligence agent sitting in a smoggy coffeehouse across Beijing from Zhongnanhai could be pumping email exchanges among the leaders, or their spouses and children, back to Washington.

The simplest way to think about the ANT catalog was that it updated the "bugs" that intelligence agents had been putting into telephones since the 1920s. But that misses the scope of what the equipment can pick up from computer networks, and the opportunities for cyberattack. The catalog revealed a new class of hardware with a scale and sophistication that enabled the NSA to get into—and alter data on—computers and networks that their operators thought were completely sealed off from the Internet, and thus impermeable to outside attack. The NSA had even gone to the trouble of setting up two data centers in China, apparently through front companies, whose main purpose was to insert the malware into computer systems.

The system, which the NSA called "Quantum," was used beyond China: there were parallel efforts to get malware into Russian military networks and systems used by the Mexican police and drug cartels.

Not surprisingly, when the *Times* prepared to publish some of these details, NSA officials declined to confirm, at least on the record, that the documents described any of their programs. Off the record, they said it was all part of a new doctrine of "active defense" against foreign cyberattacks. In short, it was aimed more at surveillance than at "computer network attack"—NSA-speak for offensive action.

The problem, of course, is that the Chinese would never believe this. When Americans find similar "implants" in our gas-distribution network, or financial markets, we immediately assume the worst—that China is preparing to attack. I asked one senior NSA official how they might signal to China, or any other adversary, that these were merely monitoring tools, not digital land mines set to explode in a few years.

"That's the problem," he said. "We can't convince them. And they can't convince us."

There was another problem. NSA officials didn't think the ANT catalog came from any documents that Snowden's "crawler" had touched. They began looking for another insider—a second Snowden.

THE SNOWDEN DISCLOSURES are now years old and describe some activities that took place before Barack Obama was elected president. As a result, some officials now argue that the damage to the NSA, while severe at the time, has diminished dramatically. Like a new iPhone or a blazing-fast laptop, the technology of surveillance and attack can look dated after a year or two. And programs that seem vital when they are created can be overtaken by events—or the arrival of new technology.

That was the argument that Adm. Rogers made to me when I first visited him at the NSA in 2014, just as he was taking over as the director of the agency and the chief of US Cyber Command. Yes, some terrorist groups had changed their tactics once they figured out how the United States was listening in on them, he said, pacing to stretch a bad back in his NSA office in Fort Meade. Yes, he acknowledged, many allies were angry—some because they discovered Washington was spying on them (like the Germans), and others because Snowden revealed they were secretly helping the Americans (the list was long, but also included the Germans).

But then he added: "You have not heard me as the director say, 'Oh, my God, the sky is falling.' I am trying to be very specific and very measured in my characterizations."

But as the discussion went on, it became clear that Rogers worried about one more long-lasting effect of the disclosures: they could, he said, take an unseen toll on the willingness of allies to work with the United States and share what they learned about the world. It was not that the intelligence agencies of Germany, France, or Britain were shocked by what they read: they knew that America spied on them, and they had plenty of programs of their own to spy on the United States. Rogers's fear was that the need for leaders in those countries to publicly condemn American overreach would have a corrosive effect on future cooperation.

Clearly, the biggest worry surrounded Chancellor Angela Merkel. One of the documents—which also now appears to have come not from the Snowden trove but from another insider leak—strongly suggested Merkel's personal cell phone had been tapped after she became the party leader of the Christian Democratic Union. That was a decade and a half earlier, long before anyone seriously thought she could emerge as leader of the country. It wasn't a cyber operation; it was plain old phone tapping.

But Merkel was outraged—and she let Obama know it. "Spying among friends—that simply isn't done," she said, at one point waving her hopelessly-out-of-date personal cell phone at reporters. Of course it was done, all the time, including by Merkel's own intelligence agency, the BND.

It was never quite proven that the NSA was actively listening to Merkel. But Obama was forced to take the unusual step of publicly declaring that he was taking his close ally off the list of NSA targets.

Unsatisfied, Merkel called Obama and, as she later said, reminded him that she had grown up in East Germany under the Stasi, the secret police. And in her arch way, Merkel made it clear to Obama that she thought there was little difference between how that secret police force monitored its people and what the United States was doing with allies. "This is like the Stasi," Merkel told him.

But Merkel was hardly the only target—the United States was also listening to the leaders of Mexico and Brazil. (After publicly declaring that Merkel would no longer be a target, Obama wouldn't say which other world leaders came off the list, and more important which ones stayed on.)

The lesson of the Merkel affair was that the NSA, in its single-minded passion to pick up every bit of foreign intelligence that it could, failed to consider the damage that might be done if its activities ever became public. No one was reviewing its target list to see if it passed the simple test applied to covert actions at the CIA: if this operation was splashed across the front pages of the *Times* and the *Post,* would someone have to resign in disgrace? In fact, a senior Obama national-security official told me that while the CIA's covert actions were reviewed every year, no one had done the equivalent on a regular basis with the NSA. That quickly changed.

Still, intelligence leaders were unapologetic, other than for getting caught. Intelligence agencies were created to spy on foreigners, they said, both friend and foe. "We're talking about a huge enterprise here, with thousands and thousands of individual requirements," General Clapper, who found himself in the crosshairs after the revelation, told members of Congress.

He justified the spying on allies with a simple mantra: trust no one. The United States spies on its friends to see "if what they're saying gels with what's actually going on," he added, and how the words and actions of other nations "impact us across a whole range of issues."

Clapper was speaking the truth. But it was a realpolitik moment because the revelations had made clear how voracious America's data appetite had become. In Germany, and around the world, the NSA was trying to gather cellular and landline phone numbers—often obtained from American diplomats—for as many foreign officials as possible. The contents of the intercepted phone calls were then stored in computer databases that could regularly be searched using keywords.

"They suck up every phone number they can in Germany," one former intelligence official told me and my colleague Mark Mazzetti. And during some fierce conversations between American and German intelligence chiefs after the Snowden revelations, the United States made clear it was not about to stop the practice, except in the case of the German chancellor's own phone.

Obama and Merkel struggled to repair the damage. "Susan Rice has been very clear to us," one senior German official told me at the time, referring to the US national security adviser. "The US is not going to set a precedent" by vowing not to spy on an allied government.

The Snowden revelations forever changed the way Germany thought about its post–World War II ally. Politicians in both Washington and Berlin like to celebrate the closeness of the alliance and describe it as unshakable. The relationship is close, but clearly shakable—and rooted in some mistrust.

Clapper insisted that Snowden merely gave Americans a vision into what Congress set up the agency to do: break into foreign signals intelligence. And so he saw Snowden as just a malicious actor, who talked about protecting Americans from snooping but revealed much more to American adversaries. "What he exposed was way beyond so-called domestic surveillance," Clapper said.

Later, after he had left office, he told me that Snowden's revelations forced the United States to end a program that had helped stop IED attacks—the improvised explosive devices that killed and maimed so many Americans and civilians alike—in Afghanistan. "The day after Glenn Greenwald wrote about it in the *Guardian*, it was shut down," he contended, referring to the American who became Snowden's biggest supporter. "He did huge damage that we're all paying for," Clapper insisted. "He was a narcissistic, self-centered ideologue."

All true. But he may have also done us a favor by forcing Washington and the new giants of the Internet—Google, Facebook, Microsoft, Intel—to rethink their relationship with the US government as well.

CHAPTER IV

MAN IN THE MIDDLE

No hard feelings, but my job is to make their job hard.

—*Eric Grosse, Google's head of security, talking about the NSA*

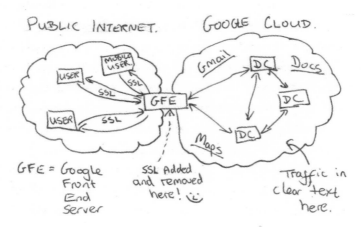

IT WAS THE smiley face that got to the engineers at Google.

The face was drawn at the bottom of a handwritten diagram on yellow paper that looked a bit like something an engineer might sketch at a coffee shop—save for the fact that it was on a slide marked TOP SECRET//SI//NOFORN and included in Snowden's trove of leaked documents.

The diagram revealed that the NSA was trying, maybe successfully, to insert itself in the nexus between the "Public Internet" and the "Google Cloud" in a move called a "man in the middle" attack. In other words, everything that went into and came out of Google's international data centers, connecting its customers around the world, could be intercepted. The drawing included an arrow pointing to the place in the diagram that corresponded to where the NSA was inserting itself. Next to the arrow, adding insult to injury, the author of the slide had doodled a smiley face.

The diagram made a single fact very stark: the NSA was working to secretly infiltrate the communications links connecting the various "front-end servers" that Google had distributed around the world, and which were used to store everything its customers held dear. Those servers were spread around the world for a very practical reason: speed of access to information. Someone in Singapore pulling documents down from the Google Cloud didn't have to wait until the data made its way halfway around the Earth from Scotland.

By finding a way in between two servers, the NSA would be able to intercept all kinds of traffic moving between them and the outside world, from Gmail messages to Google documents, even searches on Google Maps. With that one brilliant stroke of digital spy craft, the NSA would gain access to data from hundreds of millions of accounts—mostly those of non-Americans, but the accounts of millions of Americans as well. To harvest the data on "U.S. Persons," the NSA would have to get a court order, but foreigners—anyone who wasn't legally an "American person"—were fair game for the NSA, no court orders required. For the first time, the NSA would have access to the thinking, search habits, and secret communications of millions of people overseas—allies and adversaries alike. It was an intelligence agency's dream.

The diagram didn't specify exactly how the NSA was planning on getting between those servers, but there were only a few possible options. The NSA would have to hack in remotely from one of its bases around the world, physically tap the undersea cables themselves, or get cooperation from a foreign partner, such as the British. The most likely method was physically tapping the termination points in a country where the undersea cables came ashore. And since Google had not gotten around to encrypting the data that was "in transit" through these cables, merely getting into the network itself was the price of admission to the data.

When the *Washington Post* first published the slide, on October 30, 2013, over four months after the first Snowden revelations, the reaction inside the Googleplex in Mountain View was immediate and predictable.

"Fuck these guys," wrote Brandon Downey, one of Google's security engineers, on his Google Plus page, before going on, in true Silicon Valley fashion, to compare the moment to a scene from *Lord of the Rings:* "It's just a little like coming home from War with Sauron, destroying the One Ring, only to discover the NSA is on the front porch of the Shire chopping down the Party Tree and outsourcing all the hobbit farmers with half-orcs and whips."

Google's official response was only slightly more diplomatic: "We are outraged at the lengths to which the government seems to have gone to intercept data from our private fiber networks, and it underscores the need for urgent reform."

Not surprisingly, the US government was not especially interested in discussing "reform." In the NSA's view, Google's networks were fair game for message interception, just as the fiber-optic cables traversing the globe were open for intercepting message traffic. As long as the communications in question were between foreigners, and involved no "US persons" as defined by the law, intercepting their Gmail traffic and their searches was all part of a day's work for America's digital spies.

But the government, and the NSA in particular, had missed a major turn in the way Americans viewed the importance of the privacy of the data they now carried on their smartphones and laptops. When phones were landlines, hardwired to the house, and international calls

were expensive and rare for ordinary Americans, there was little public outrage if the government kept tabs on international phone lines.
And in the years after the September 11 attacks, there was considerable
public sympathy for the government's interest in going after terrorist
communications.

All that changed with the invention of the smartphone. Suddenly,
the information the NSA was sweeping up wasn't just telephone traffic. For the first time people were keeping their whole lives in their
pockets—their medical data, their banking information and work
emails, their texts with spouses, lovers, and friends. It was all being
stored in those Google servers, and others like it run by Yahoo! and
Microsoft and smaller competitors. And depending on where one was,
that data could be stored anywhere. The distinction between "international" communications and "domestic" communications was virtually wiped out. All of a sudden the idea of the government's getting
inside Google's servers seemed a lot more worrisome.

The smiley-face revelation deeply threatened Google's business
around the world—along with that of Facebook, Apple, and every
other corporate symbol of America's newest form of soft power. The
note suggested Google was being unwittingly hacked by the US government, but in a world where trust in institutions is in increasingly
short supply, Google and the other tech giants couldn't help but appear
complicit. Customers in Germany or Japan would suspect that American companies were secretly turning over their data to the NSA. (That
assumption was not entirely wrong: if served an order from the Foreign
Intelligence Surveillance Court, Google and the other big communications providers had no choice but to comply.) And governments around
the world could use these Snowden revelations to make the case that
the American firms were inherently untrustworthy and should be regu-
lated or barred from their countries.

"We are not in a good place," Eric Schmidt, Google's then chief
executive told me soon after the Snowden disclosures. But as Schmidt
himself conceded, Google's relationship with the US government was

complicated—a lot more complicated than Brandon Downey's outburst might suggest.

THE BURDEN OF solving the immediate problem—locking up Google's systems and convincing customers around the world that their data were not being piped straight to Fort Meade—fell to Eric Grosse, the amiable head of Google's security efforts. With pale eyes and gradually graying hair, Grosse looked more like a bespectacled suburban dad than one of Silicon Valley's key players, but his Stanford computer science PhD and hobby of piloting his own high-winged plane, gave him Valley credibility.

By the time my colleague Nicole Perlroth and I arrived at Grosse's office on the edge of the Google campus in early summer 2014, he was deeply engaged in figuring out how to block every pathway the NSA could carve into the tech giant's networks. In the open cubicles that looked out on the old air base, Moffett Field, in the heart of what is now Silicon Valley—the remnant of a pre–World War II age when the airplane was remaking global power—Grosse and his team of engineers were working day and night to NSA-proof the Google systems.

The project, Grosse said, had actually begun long before they saw the smiley-face diagram. As early as 2008, Google had been investing in consortiums that laid undersea cables. But sharing had its risks: The company was not fully in control of who else had access to the cables, and there was always the risk that Google's traffic could be thrown off the lines during an emergency, leaving its users without access to their data. Less than a decade later, Google had undertaken a multibillion-dollar effort to put down its own fiber-optic lines across the Atlantic and the Pacific so that it could control the speed and reliability of information that flowed between its servers and its users.

In addition to laying its own cables, Google had also decided, before the Snowden revelations, to roll out a program to encrypt all the data that ran between its data centers. But as with the cable laying, the encryption effort was still plodding along when the smiley-face document made it clear that the US government was intent on breaking into Google's networks. Suddenly, making sure no one else was inside those networks became an urgent priority.

The possibility that an intelligence service could drill into Google's networks had always been a concern, Grosse told us, but until this document surfaced the threat had seemed theoretical. "Reasonable people could differ on the risk," he said. The prevailing view among his colleagues before Snowden was that getting into the communications lines between Google servers "would be too costly," even for the NSA. "Most of the traffic was not sensitive, so huge processing would have to be done to mine any nuggets," he explained. Who in their right mind would tackle a haystack that big in order to find a few needles?

Yet there were a few security engineers both inside and outside Google, Grosse admitted, who better understood the mind-set of the NSA and thought the agency might be motivated to tackle the haystack. "They had read about Operation Ivy Bells," he said.

Grosse was referring to the huge US intelligence project in the early 1970s to tap into the Soviet Navy's undersea cables in the Sea of Okhotsk. At significant risk of discovery, the NSA had dispatched a submarine to wrap a secretly developed twenty-foot-long set of devices around the cables to record all the message traffic. Every month or so divers would slip into the waters, descend four hundred feet, and retrieve the recordings. The operation ran with great success until 1980, when a forty-four-year-old NSA communications specialist with a personal bankruptcy problem walked into the Soviet embassy in Washington and blew its cover.

During its years in operation, Ivy Bells was a complicated operation with significant technical challenges. Overcoming those challenges took help from the tech giant of the time, AT&T's Bell Laboratories, which considered itself not just an American firm but also a founding member of the intelligence-industrial complex that ultimately won the Cold War. There was no shame in the revelation of the company's role; the undersea cable in question was used by the Soviet military to communicate with subs carrying ICBMs aimed at American cities. Bell Labs had done what many Americans expected they would do.

But the world had changed in the four decades between the start of Ivy Bells and the operation described in the smiley-face drawing. Google engineers, by and large, hadn't grown up in a Cold War environment and did not view their role as one of supporting American defense and intelligence partners. In any case, this wasn't a collaborative effort; the government was breaking into a company's network and helping itself to the information. And although Google was born an American company, it certainly saw itself as much more of a global citizen than Bell Labs had in its heyday. For the Google engineers sitting in the cubicles surrounding Grosse, their number-one mission was not to uphold American national security but to assure their customers that their information was safe. That included Google's global customers, not just Americans.

All of a sudden, Grosse recalled, Google's sense of risk had changed. The NSA had exploited "the last chink in our armor," he told us as we walked around the Google offices, where some programmers had put up pictures of the NSA headquarters building with slashes through them. The encryption project, "originally operating on a timescale of months," now had to be "finished on a timescale of weeks."

He put the program into overdrive. This was just one more skirmish, he understood, in what would be a long-running war between Silicon Valley and the NSA. "No hard feelings," Grosse said to us as we walked through the security center Google had erected to seal up its vulnerabilities, "but my job is to make their job hard."

Google soon added a new email-encryption feature to its product line. And, in a pointed jab at the NSA, the code ended with a smiley face.

· · ·

THE SNOWDEN AFFAIR kicked off a remarkable era in which American firms, for the first time in post–World War II history, broadly refused to cooperate with the American government. They wrapped some of that refusal in Silicon Valley's typical libertarian ideology. But their real fear was that any open association with the NSA would prompt customers to wonder whether Washington had bored holes into their products.

Snowden gave allies and adversaries alike grounds for an argument about the dangers of using American technology—the same argument the United States raised regularly about letting Kaspersky antivirus software, designed in Russia, run on American computers, or permitting Chinese firms to sell their cell phones and network equipment in the United States. And the Snowden revelations set the stage for massive conflicts between the powerhouses in the technology world— Google, Apple, Facebook, and Microsoft—and a National Security Agency that had blithely assumed American companies would be on its side, just as Lockheed and Boeing and Raytheon had been during the Cold War.

Google was hardly the only target of the NSA, and the project to get inside its servers was a small part of a far broader effort. Months before the Google smiley face emerged, Snowden leaks had revealed the existence of the NSA program code named "PRISM," in which the NSA siphoned off Internet communications of all types under orders issued by the Foreign Intelligence Surveillance Court. The companies were ordered to stay quiet about the program, and they were paid several million dollars to compensate them for the cost of compliance. Inside the NSA the operation was run by a group known as "Special Source Operations," which sought to recruit American companies to the cause after the September 11 attacks. Everyone from Microsoft to Yahoo! to Apple to Skype participated, some more reluctantly than others. Government intelligence analysts working around any encryption systems could search the companies' huge databases of information.

The smiley-face document, combined with the PRISM revelations, suggested the NSA was actually accessing corporate servers two different ways: the court-ordered way, with all its legal oversight, and via a covert effort overseas, for which no court order was needed but much stealth was required.

That distinction was lost on most of the world. To anyone who wasn't looking deeply, it appeared that Silicon Valley companies were simply opening their doors to the NSA—that they had become an arm of an overbearing US government. The tech companies were quickly forced to make statements distancing themselves. On June 7, the day the *Guardian* revealed the PRISM program's access to Facebook, Microsoft, and others, Mark Zuckerberg posted a heated defense: "Facebook is not and has never been part of any program to give the US or any other government direct access to our servers," he asserted. That same day, Microsoft declared that it complied only "with orders for requests about specific accounts or identifiers. If the government has a broader voluntary national security program to gather customer data, we don't participate in it."

It turned out, though, that other companies did participate in such a program. Snowden documents showed a long history of cooperation between the NSA and the telecommunications giants who controlled what had become the backbone of the Internet. Documents in the trove showed that seventeen AT&T Internet hubs in the United States had installed NSA surveillance equipment; a smaller number of Verizon facilities were similarly equipped. Often, AT&T engineers were among the first to test the government's new technology. One internal NSA memo, recording a trip that one of the agency's senior officials took to the company, referred to the firm's "extreme willingness to help with SIGINT and Cyber missions and the breadth and depth of not only the program's access" but the "amazing knowledge" of the company's workforce.

It was a relationship so vital that President Obama sometimes intervened directly with telecommunications executives, calling them personally to ask for help if there was a critical need for information about a terrorist group or if intelligence agencies feared that an attack could be imminent. And the executives, while on the line with the president of the United States, had to essentially ask themselves a hard question: Were they an American company first, or a global one?

"It was the kind of thing you couldn't say no to," one chief executive said to me. "You have a president saying lives are on the line."

But now that chief executive also had to contemplate the dangers of saying yes. After Snowden, the potential cost of cooperating with Washington was a lot higher. Any country that wanted to keep American firms out of their markets could make an easy national-security argument: buy the American equipment, and you were probably buying a "back door" that the NSA installed to tap into those systems.

In the year after the Snowden revelations, I spoke with a number of American executives who crisscrossed Europe and Asia arguing to customers that this wasn't the case. But they found little traction. "We assume that if a company is based in the United States, there is an understanding with the American intelligence services," a senior German intelligence official told me one night over dinner in Berlin. "And look at who is supplying the CIA!"

He was referring to Amazon's $600 million deal to build the CIA's enormously complex "cloud storage" system. For the agency, the deal was a true revolution—a response to the critique that information in the intelligence agencies was so stuck in "silos" that it was impossible to conduct the kind of data-crunching analysis that would reveal patterns, or plotters. Gradually, the Pentagon started to move in the same direction. Of course, the whole idea of centralizing information in the cloud made many senior American officials exceedingly nervous: If the lesson of 9/11 was that more information needed to be shared, the lesson of Snowden was that centralized systems can yield huge leaks. But the real irony was that as tech firms publicly protested the NSA's intrusions into their networks—Microsoft, IBM, and AT&T among them—they were privately vying for the hugely profitable business of managing the intelligence community's data.

.   .   .

TIM COOK, THE quiet, almost ascetic chief executive of Apple, rose in the company as a counterweight to Steve Jobs. Jobs was the showman, Cook the understated strategist. Jobs erupted when products didn't look right or politics limited the ideal technological solution; Cook lacked Jobs's intuitive sense of what made a product feel distinctly like an Apple product, but what he lacked as a designer he made up for with his considerable geopolitical sensibility. Whereas Jobs was no ideologue and rarely dug deeply into the question of Apple's place in society, Cook seemed as comfortable making a civil-liberties argument as a technological one.

Perhaps Cook's social and political intuition was the result of his years growing up in Alabama, where one of his searing memories was of bicycling by a group of Klansmen burning a cross on the lawn of a black family in the town of Robertsdale and yelling at them to stop. "This image was permanently imprinted in my brain and it would change my life forever," he said in 2013, the same year as the Snowden revelations.

Cook spoke little about his personal life; it was not until he was already running Apple that he began to talk about growing up as a gay teenager in a conservative state. He kept portraits of Robert Kennedy and Martin Luther King Jr.—two heroes from his youth—in his office. So when Obama invited him into the White House—along with Vinton Cerf, one of the founders of the Internet, and Randall Stephenson, the chief executive of AT&T—in late 2013, in an attempt to contain the damage of Snowden's recent document dump, Cook certainly had views. He already knew where he was headed with Apple's technology—in exactly the direction that the government did not want him to go.

At the heart of Cook's dispute with the government was whether it was more important for Apple to secure the data that users keep on their phones, or to assure the FBI and the nation's intelligence agencies that they could get inside any iPhone. For Cook, this was not a moral dilemma, and it was an even easier business question. He had risen to the top at Apple talking about how one of the company's fundamental goals was to help Apple users keep their digital lives private. Apple made its money off of hardware and apps, not the ability to sell ads around email services or search engines.

But to Cook's surprise, he ran into a buzz saw of opposition, not only from the FBI but from Obama himself. The latter was surprising because Cook was one of just a few tech executives with whom Obama had worked up something of a friendship. The two men were about the same age, and on visits to Washington—which Cook made much more frequently than Jobs had done—Cook would sometimes slip into the White House for a quiet meeting with the president, who admitted to something of a fascination with Apple. (His aides guessed that this was because the NSA and the White House Communications Agency insisted they could not make the iPhone secure enough for him to use and left him with a BlackBerry, which he hated.) For his part, Obama seemed in his element on trips to Silicon Valley, where he could wax freely about the future of the American economy and know that he was among his core supporters. Their relationship made the open conflict with Cook, and Obama's other tech supporters, over encryption all the more fascinating.

The Snowden revelations forced Cook to take a stand in a battle that had been brewing for years—the FBI's fight against the growing movement toward personal encryption. Everyone agreed that bank information and certainly classified government files should be encrypted. But the idea that the same could or should be true of every individual's personal data was relatively new, and it was one that chilled federal law enforcement. The FBI warned that encrypted personal communications were creating a "going-dark" crisis that would keep its agents—along with local police—from tracking terrorists, kidnappers, and spies.

No one embraced this view more fully than James Comey, Obama's FBI director, who had come to public prominence standing up to the Bush White House when it sought to route around the law and the courts in authorizing a wiretapping program. But Comey was a government lawyer at heart, and in this case he argued that without a court-approved way into Apple's phones, what most people called a "back door," the main appliance of our lives was giving ISIS plotters and homegrown terrorists a cheap, secret way to communicate.

That argument struck some as disingenuous. Certainly encrypted communications made it hard to intercept conversations that had previously taken place in the clear. But this ignored the flood of new, Internet-enabled technologies that had given rise to—as more than a few technologists noted—the "golden age of surveillance." In a world in which one's car and lost luggage could be tracked electronically, where a Fitbit broadcasts the wearer's location and people's watches are connected to the Internet—life is a lot easier for investigators. As one FBI investigator admitted to me: "If you put us in a time machine and took us back ten years, we'd feel like our best tools have been taken away."

Cook believed that giving the US government the back door it demanded into his products would be a disaster. "The most intimate details of your life are on this phone," he said to me one day in San Francisco, holding up an iPhone. "Your medical records. Your messages to your spouse. Your whereabouts every hour of the day. That information is yours. And my job is to make sure it remains yours."

And so a year later, in 2014, Cook went to war with the Obama administration over encryption.

WHEN COOK TOOK the stage in Cupertino in September of that year to announce the iPhone 6, Apple advertised it—not in so many words—as the phone for the post-Snowden age. Over the years, even before the Snowden revelation, Apple had gradually encrypted more and more data on its phones. Now thanks to a software change, the phone would automatically encrypt emails, photos, and contacts based on a complex mathematical algorithm that used a code created by, and unique to, the phone's user.

But the bigger news was that Apple would not hold the keys: those would be created and held by users. That change marked a huge break from the past. Until then, Apple always had the keys, unless someone using one of their phones was making use of a special encryption app—a complicated process for most users.

Now encryption would become automatic. "We won't keep those messages and we wouldn't be able to read them," Cook told me. "We don't think people want us to have that right." Even worse for those who might need to get into an iPhone, breaking any individual code would take awhile: the Apple technical guide reported that it could take "more than 5½ years to try all combinations of a six-character alphanumeric passcode with lowercase letters and numbers."

Over the next week or so, the implications soaked in at FBI headquarters. If encryption were automatic, it would be nearly universal. And so if the FBI went to court and delivered an iPhone to Apple and demanded its contents, even with a valid court order taped to it, they would get back a pile of encrypted gibberish. Apple would argue that without the user's code, the company had no way of decrypting the information. If the government wanted the data, it had better start trying to break the code by brute force.

Comey hadn't seen a move of this scope coming. It was a complete breach of the tacit understanding he had grown up with, enshrined during the simpler days of the Cold War and the shock of 9/11, that there would always be a way around encryption. Suddenly, that assumption was changing. "What concerns me about this is companies marketing something expressly to allow people to hold themselves beyond the law," Comey said after the announcement at a news conference devoted largely to combating terror threats from the Islamic State.

To those who had been around since the "Crypto Wars" of the 1970s and '80s, it was a familiar argument. At that time, the NSA wanted to control all cryptography research so that it could read anything it wanted. It had fought academics and private firms that wanted to publish cutting-edge research into the most secure ways of encrypting data, and it wanted a role in setting the standards for cryptography so that it could read messages sent around the world. In short, the NSA wanted to control the development of cryptography so that it wouldn't be locked out of any system.

Then, in the 1990s, the agency developed the "Clipper chip," a chip that could be installed in computers, TVs, and early cell phones. The Clipper encrypted voice and data messages but had a back door that the agency could unlock, ensuring that with proper authorization the intelligence agencies, the FBI, and local police could decode any message. The Clinton administration endorsed the idea—for a while—arguing that once the chip went into every device, there would be no way for terrorists to avoid using it and the intelligence agencies could listen in.

Naturally, consumers and most manufacturers rebelled, and the Clinton administration retreated. "NSA lost both battles," noted Susan Landau, an expert on the history of these conflicts.

Now the fight was being waged once more, and with a startling ferocity. Snowden's revelations made tech companies more determined than ever to beef up their encryption, their case made easier as consumers read about countless hacks on companies that stored their credit-card data. Obama read the tea leaves and created an independent panel to advise him on what kind of new restrictions—if any—to put on the NSA after the Snowden revelations and to guide him on balancing privacy and security. The panel included Mike Morell, who had retired from the CIA not long after I dealt with him on Olympic Games, and a range of other former counterterrorism officials, academics, and constitutional lawyers.

To the shock of the NSA and the FBI, Morell and his colleagues sided with Big Tech. The panel made a unanimous recommendation that the government should "not in any way subvert, undermine, weaken, or make vulnerable generally available commercial software." Instead, it should "increase the use of encryption and urge US companies to do so."

No sooner had the ink dried on the panelists' signatures than the

NSA urged Obama to ignore its advice. With terror groups already turning to encrypted apps with names like Telegram and Signal, Landau noted, "the last thing the NSA wanted was to make encryption easier for everyone across the world; how would it listen in abroad?"

The burden of arguing for government access fell on Comey, who had a bit of a flair for the dramatic—as the world later learned in his confrontations with Hillary Clinton and Donald Trump—and soon he was reaching for the most emotive example he could come up with to support his position: What would happen, he asked, when the parents of a kidnapped child came to him with a phone that might reveal the whereabouts of their kid, but its contents could not be determined because they were automatically encrypted—just so that Apple could extend its profits around the world? Comey predicted there would be a moment, soon, when those parents would come to him "with tears in their eyes, look at me, and say, 'What do you mean you can't'" decode the phone's contents?

"The notion that someone would market a closet that could never be opened—even if it involves a case involving a child kidnapper and a court order—to me does not make any sense," he said. He extended the analogy to apartment doors and car trunks to which there were no keys. That would stymie a legal search warrant, he said. If it wouldn't be tolerated in the physical world, he said, why should it be tolerated in the digital world?

From the other coast, Tim Cook had an answer: the apartment keys and trunk keys belonged to the owner of the apartment and the car, not to the manufacturer of their locks. "It's our job to provide you with the tools to lock up your stuff," Cook said. At Apple and Google, company executives told me that Washington had brought these changes on themselves. Because the NSA had failed to police their own insiders, the world was demanding that Apple prove their data was secure, and it was up to Apple to do so. Naturally the government saw this as a deliberate dodge. And to some extent it was.

But Cook had a bigger and better argument, one that the government could not so easily parry: if Apple created a back door into its code, that vulnerability would become the target of every hacker on Earth. The FBI was naïve to think that if the tech companies created a lock and gave the FBI a key, no one else would figure out how to pick it. "The problem is," Cook said, "anyone with any technical skills knows that if you create an opening for the FBI, you create one for China and Russia and everyone else."

Discreetly, Cook took that argument to Obama himself—in quiet sessions in Washington and Silicon Valley. American spy agencies and police had all kinds of other options, he argued. They could find data in the cloud. They could use Facebook to figure out anyone's acquaintances. But to give them access to that data inside the phone was to undercut an American expectation of privacy—and to invite the Chinese and others to do the same, for far more nefarious purposes.

"The only way I can protect hundreds of millions of people is the way I'm doing it," Cook told me during one of his Washington visits, fresh from making this case to Obama and his aides. But he knew that despite his admiration for Obama—"I love the guy," he'd often repeat—he was losing the argument with him.

Obama was looking to straddle the problem by arguing that security and privacy could be balanced. In Cook's view, that slogan sounded nice from the White House Press Room but made no technological sense. Drilling a hole in the iPhone operating system was like drilling a hole in your windshield; it weakened the whole structure and allowed everything to fly in.

When I spoke with Cook, it was clear that he was worried about another problem, one that American officials weren't discussing in public because it so complicated their case. China was watching Apple's struggle with the US government—and it was rooting for Comey. If Apple agreed to create a back door for the FBI, China's Ministry of State Security would give Apple no choice but to create one for them too, or else be ejected from the Chinese market.

At the White House, many officials worried about being accused of becoming an accessory to China's growing crackdowns on dissidents. In fact, the fear paralyzed some of them. But FBI officials quickly waved away this argument. "We're not the State Department," one of Comey's top aides told me. The rest of the intelligence community seemed likewise unconcerned. Just days after the Apple announcement, the director of one of America's sixteen intelligence agencies invited me to his office to rail against Apple's top executives.

"This is a direct result of Snowden," he declared, the only thing on which he and executives like Cook seemed to agree. "We're going blind." Smartphones, he said, were routinely part of "the pocket litter" of every terrorist tracked down by American Special Forces in Pakistan and Afghanistan, and now in the Islamic State. Most were drained of data on the spot. Now they would routinely carry levels of encryption previously available only to government agents of Russia and China.

"It's a terrible choice," another spy chief told me. "We have to decide whether to attack our own companies" or live in a world in which the working assumption of the Western intelligence agencies—that they could obtain any message, break any code—would no longer apply.

The battle lines were drawn. But the big fight would not come for another year, just as the presidential race of 2016 was heating up.

JUST BEFORE MIDDAY on Wednesday, December 2, 2015, Syed Rizwan Farook and Tashfeen Malik armed themselves with assault rifles and semiautomatic pistols and attacked a holiday party at the city's health department in San Bernardino, California. Fleeing the site, they left behind an explosive constructed of three pipe bombs that failed to detonate. Fourteen people were killed and twenty-two were injured. The dead ranged in age from a twenty-six-year-old woman who was raising a toddler to a sixty-year-old Eritrean immigrant who had left for a safer life in the States and raised three children with his wife.

The attackers were killed in a shootout a few hours later, about two miles from the site of the shootings. Farook, twenty-seven, turned out to be the son of Pakistani parents who had immigrated to Illinois before he was born, making him a natural-born American citizen. Malik, a year his senior, was born in Pakistan but lived in Saudi Arabia with family—before she came to the United States after meeting Farook on his hajj, or pilgrimage, to Mecca. Radicalization followed. It turned out she had pledged her allegiance to ISIS on Facebook just before the attack. But no one had noticed until it was all over.

Then came the detail that would reignite the encryption debate for months. Farook had left behind his work-issued iPhone 5c. It was critical, because while the couple had worked hard to cover their electronic tracks prior to the attack—smashing personal phones and hard drives, deleting emails, and using a disposable burner phone—they had forgotten the work iPhone. The FBI believed this device would provide vital evidence: Farook's communications with any associates and, most vitally, his GPS coordinates just before the attack. That is, if they could get past the phone's encryption. (Farook did not upload his data to iCloud, which would have been more accessible.)

The problem was that Farook had locked his phone with a code, and now he was dead. And while the FBI could try a brute-force cracking—essentially trying all possible combinations—Apple's safety features include one that wiped all data after ten wrong password attempts. That feature was designed to protect users against any hacker who broke into a phone—mostly criminals seeking financial information, credit-card numbers, or information about how to gain access to a house or a safe.

For Comey, here was a case tailor-made to fit his argument. If there were other ISIS-inspired Americans or immigrants who were in communication with Farook and Malik, they needed to be picked up quickly. And Apple, in the name of privacy and security, was arguing that it didn't know Farook's passcode, so it could be of no help. Comey publicly asked Apple to write new code—essentially a variant of the iPhone operating system—that would allow the FBI access to iPhone password-protection security features, thereby circumventing the problem and gaining access to Farook's phone. Comey insisted he would use the new code with discretion. In fact, the FBI may have already possessed the technology to unlock the phone, according to a subsequent report by the FBI inspector general. But investigators were told that technology was available mostly for foreign intelligence work, and the inspector general concluded that senior FBI officials were looking forward to a court confrontation with Apple.

The Justice Department got a federal magistrate in California to order Apple to find a way to crack the phone. Cook realized immediately that Comey saw the San Bernardino case as a chance to short-circuit the intensifying arguments about encryption and had escalated the dispute to the courts. Cook saw this moment as a chance to take a stand and to show his independence from the FBI. He wrote a 1,100-word letter to his customers that was striking for its accusation that the Obama administration was so obsessed with access that it was ready to sacrifice the privacy of its citizens.

> The United States government has demanded that Apple take an unprecedented step which threatens the security of our customers. We oppose this order, which has implications far beyond the legal case at hand. . . . Some would argue that building a backdoor for just one iPhone is a simple, clean-cut solution. But it ignores both the basics of digital security and the significance of what the government is demanding in this case. . . . The implications of the government's demands are chilling . . . ultimately, we fear that this demand would undermine the very freedoms and liberty our government is meant to protect.

For the leader of one of the most successful companies on the planet, larger than some European economies, it was a remarkable accusation. Cook was charging an administration that treasured its reputation as a progressive force for civil rights with seeking to undermine a core constitutional principle about individual freedom. With Apple and the FBI at a standoff, Obama dispatched some of his senior intelligence officials to Silicon Valley to talk Cook off the ceiling and look anew for a compromise. Cook wasn't interested. Though he could not yet reveal it publicly, the FBI's demand that Apple break into the San Bernardino phone was just one of four thousand law-enforcement requests to the company in the second half of 2015.

Comey wasn't about to back down; he told aides that the publicity around the San Bernardino case would, if anything, remind criminals, child pornographers, and terrorists to use encryption. This was the moment to settle the encryption wars, he said, once and for all.

IT DIDN'T TURN out that way. In the end the FBI paid at least $1.3 million to a firm it would not name—believed to be an Israeli company—to hack into the phone. The FBI refused to say what the technical solution was, or to share it with Apple, apparently for fear that the company would seal up whatever hardware or software loophole was discovered by the hired hackers. Later the FBI told Congress they didn't actually know what technology was used: They hired a locksmith, who picked the lock. They deliberately didn't ask how it was done—because the White House, under its own rules about disclosing most vulnerabilities to manufacturers, might have been forced to clue in Apple.

Obama, the constitutional law professor, never solved this problem. And he never implemented the recommendation from his own advisory panel that the government encourage the use of more and more encryption. He told his aides that years of daily warnings in the President's Daily Brief about terrorist activity around the world had altered his view: The United States simply could not agree to any rules that locked it out of some conversations. It was a breach with the tech community that he simply never overcame.

"If, technologically, it is possible to make an impenetrable device or system, where the encryption is so strong that there is no key, there is no door at all, then how do we apprehend the child pornographer?" Obama asked publicly a few years later. "How do we disrupt a terrorist plot?"

If the government cannot crack a smartphone, Obama concluded, "then everyone is walking around with a Swiss bank account in your pocket."

Obama had accurately described—but hadn't solved—one of the central dilemmas of the cyber age.

CHAPTER V

THE CHINA RULES

I mean, there are two kinds of big companies in the United States.
There are those who've been hacked by the Chinese and those
who don't know they've been hacked by the Chinese.

—*James Comey, then FBI director, October 5, 2014*

THE BOXY TWELVE-STORY building along Datong Road on the outskirts of Shanghai was easy to overlook. In the jumble of a city of
24 million people—China's most populous, and among its most high-
tech—it was just another bland, white high-rise. The only hint that
the unmarked building was actually a base for the People's Liberation
Army and its pioneering cyber force, Unit 61398, came if you looked
at the protections surrounding the tower—or the security forces who
came after you if you dared to take a picture of it.

The digital addresses of many of the hackers stealing terabytes of
data from American corporations—everything from the designs of the
F-35 aircraft to the technology of gas pipelines, from data collected
by health-care systems to Google's algorithms and Facebook's magic
formula—pointed straight back to Pudong, the run-down neighborhood of massage parlors and noodle joints surrounding the white
building.

But the trail of evidence fizzled out there, at the level of the neighborhood. The Chinese had so clouded the final termination addresses of the hackers' systems that it seemed impossible to trace the thefts back to any one building. That was driving Kevin Mandia, a wry former air force intelligence officer who was leading one of the several private investigations into Chinese intrusions, absolutely crazy. It seemed impossible that the hacks he was tracing came from anywhere but the highly defended high-rise. He just couldn't prove it. Yet.

"Either these hackers are hanging out in the noodle shops and the massage places or they are working day and night in that building on Datong Road," said Mandia late one day near his office in Alexandria, Virginia.

While Mandia was building a client base of more than a hundred companies and revenues of $100 million for his cybersecurity firm, Mandiant, he had been tracking a Chinese hacking group with clear ties to the PLA. Mandiant called the group "Advanced Persistent Threat 1 (APT1)," the awkward term the industry uses to identify and number malicious state actors in cyberspace that aren't going away.

Mandia was certain the hackers were part of Unit 61398, but he also knew that accusing the Chinese military directly would constitute a huge step for his company. Over seven years, he had compiled a list of the unit's suspected attacks on 141 companies across nearly two dozen industries, but he needed solid evidence before he could name them. Yet as long as none of his investigators could get inside the building, whether physically or virtually, to identify the thieves, the Chinese would keep denying that their military had been tasked with stealing technology for state-run Chinese firms.

Ever resourceful, Mandia's staff of former intelligence officers and cyber experts tried a different method of proving their case. They might not be able to track the IP addresses to the Datong Road high-rise itself, but they could actually look inside the room where the hacks originated. As soon as they detected Chinese hackers breaking into the private networks of some of their clients—mostly Fortune 500 companies—Mandia's investigators reached back through the network to activate the cameras on the hackers' own laptops. They could see their keystrokes while actually watching them at their desks.

The hackers, just about all of them male and most in their mid-twenties, carried on like a lot of young guys around the world. They showed up at work about eight-thirty a.m. Shanghai time, checked a few sports scores, emailed their girlfriends, and occasionally watched porn. Then, when the clock struck nine, they started methodically breaking into computer systems around the world, banging on the keyboards until a lunch break gave them a moment to go back to the scores, the girlfriends, and the porn.

One day I sat next to some of Mandia's team, watching the Unit 61398 hacking corps at work; it was a remarkable sight. My previous mental image of PLA officers was a bunch of stiff old generals sitting around in uniforms with epaulets, reminiscing about the glory days with Mao. But these guys were wearing leather jackets or just under-shirts, and probably saw Mao only if they visited his mausoleum in Tiananmen Square. "They were such bros," Andrew Schwartz, one of Mandia's communications specialists, recalled later. "But they were prodigious thieves." They were also thieves with multiple employers: some moonlighted as hackers for Chinese companies, making it un-clear whether they were stealing on government or corporate orders.

This was what the new cold war between the world's two largest economies looked like up close. It bore no resemblance to the more familiar conflicts of past decades: No one was arguing over the fate of Taiwan, or bombarding the tiny islands of Quemoy and Matsu, as Mao did in 1958, prompting the United States to reinforce its Seventh Fleet and consider whether it was worth going to war. For while China was still interested in staking its territorial claims—starting in the South China Sea—and keeping America at bay, it understood the keys to reemerging as a global power after a centuries-long hiatus: artificial intelligence, space technology, communications, and the crunching of big data. And of course, outmaneuvering its only real challenger, the United States.

Yet in Washington, three American presidents—Clinton, Bush, and Obama—had struggled to define exactly what China was in relation to the United States: A potential adversary? A sometime partner? A vital market for American goods? A growing investor in the United States? China was all of these, and more, which is what made it such an intractable and fascinating foreign-policy problem. Every time the White House considered calling the Chinese out for their thefts, there was the temptation to pull its punches. There were always countervailing interests: the State Department needed help on North Korea, the Treasury didn't want to upset the bond markets, the markets didn't want to see a trade war started. In the cyber realm, this meant holding back on naming the Chinese when they got caught in some of the biggest hacks in recent years.

Instead, objections would be raised with the Chinese in closed sessions at the annual "Strategic and Economic Dialogue," assuring that any discussion would remain quiet. And they would almost always result in a scripted Chinese response: It's not us, the officials would insist. It's a bunch of teenagers, or criminals, or miscreants.

Even in 2013, as a frustrated President Obama prepared to sign a new executive order to bolster America's response to cyber intrusions, he couldn't quite bring himself to name the Chinese government as the chief offender. "We know hackers steal people's identities and infiltrate private emails," he said in his State of the Union address that year. "We know foreign countries and companies swipe our corporate secrets. Now our enemies are also seeking the ability to sabotage our power grid, our financial institutions, our air-traffic-control systems. We cannot look back years from now and wonder why we did nothing in the face of real threats to our security and our economy."

Mandia was determined to do what the government wouldn't do: publicly prove that the PLA was involved. He had come to the *New York Times,* over the objections of some of his colleagues, because he knew that an independent assessment of Mandiant's work would give it more credibility. But his real goal seemed to be to goad the US government, and private industry, into action.

"I'm not sure this is the smartest thing to do," he told me. "You know what the Chinese will do: They'll put a big bull's-eye on my back."

But Mandia didn't seem all that worried. The Chinese were different from the Russians. "They are into this stuff for the money, the technology, the military power," he said to me. "You don't see Chinese shutting down networks, though if we ever got into a war they certainly know how to do that. For them it's pretty simple: They want control at home, and access to all the technology they can eat here."

No one had expected the digital revolution in China to unfold in quite this way. In the 1990s, in the wake of Tiananmen Square and the subsequent government crackdown, it was an article of faith in Washington that the Internet would change China more than China would change the Internet. No one believed this more fervently than Bill Clinton. During a presidential visit to Beijing in 1998, he told students at Beijing University that the digital revolution meant one thing for them: more democracy, albeit with Chinese characteristics.

"Over the past four days, I have seen freedom in many manifestations in China," he told the students, who sat there likely trying to figure out what the cost might be of applauding. "I have visited a village that chose its own leaders in free elections. I have also seen the cell phones, the video players, the fax machines carrying ideas, information and images from all over the world. I've heard people speak their minds. . . . In all these ways I felt a steady breeze of freedom."

Then came the core of the argument he had practiced on me and some other reporters at the White House before he left on the trip: "The freest possible flow of information, ideas, and opinions, and a greater respect for divergent political and religious convictions, will actually breed strength and stability going forward."

Clinton told me after the trip that he emerged convinced that as China grew more connected, the Communist Party would weaken. He wasn't alone. The writer and pro-democracy dissident Liu Xiaobo, who was in prison yet again during Clinton's visit, later wrote that the Internet was "God's gift to China."

The Chinese president at the time, Jiang Zemin, listened to Clinton and clearly did not buy a word of it. Already Chinese leaders were thinking about how the West's invention could be used as an instrument for social control at home and economic advantage abroad.

As the Chinese government became increasingly skilled at using cyber technology as a tool for both domestic surveillance and coercion, American companies looked the other way. For a while Westerners seemed willing to convince themselves that although the Chinese government cracked down on Chinese users of the Internet, it would let Westerners alone, as long as they were in their protected enclaves.

But the course was set. Every year, the Chinese imposed stricter requirements to ensure that their internal security forces knew exactly who was on the Chinese Internet, and what they were saying. Officials required that users employ their real names, not pseudonyms, and eventually told Internet companies that they would have to keep all servers handling Chinese traffic physically located within China. As the restrictions tightened, Western news organizations faced an inevitable choice: play by China's rules, including complying with its expanding censorship requirements, or gradually get edged out of the world's largest market. Bloomberg, among others, folded and agreed to censor.

Over the next few years, in many different forms, this drama would play out over and over, with Facebook and Uber, Apple and Microsoft. Each would have to make its peace with the China Rules: give the state access to your company's information, and often your underlying technology, or get out.

Among the first to confront the problem was Google, whose experience taught every American company that China wasn't hacking just for hacking's sake: it had an intelligence angle and a political agenda.

. . .

As it turned out, an uncensored Google made the leadership of the country very nervous. As American intelligence agencies later learned, the leadership were Googling themselves, and the results were not always complimentary.

A secret State Department cable, written on May 18, 2009, and made public the next year in the WikiLeaks trove that Chelsea Manning had taken, reported that Li Changchun, who headed the propaganda department for the Chinese Communist Party and was a top member of the leadership, was astounded to discover that when he typed his name into a Google search bar he found "results critical of him." Since he was the government's leading censor, the fact that any Chinese citizen with an Internet connection could read something unpleasant about how he performed his duties was a rude awakening. From that moment, the die was cast.

Google's problems accelerated beyond search results. Beijing officials didn't like Google Earth, the satellite mapping software, because it showed "images of China's military, nuclear, space, energy, and other sensitive government agency installations." Knowing that George Bush was steadily pressing China to do more to combat terrorism, officials told the American embassy that Google Earth was a terrorist's best tool.

Li required the three state-owned Chinese telecommunications firms to cut off Google, preventing it from reaching hundreds of millions of Chinese users. He wanted to sever the link between Google's Chinese site, which complied with China's censorship rules—no Tiananmen Square history, nothing on the Falun Gong—and Google's Hong Kong and US sites, which had no censorship.

But in December 2009, Google's top executives discovered a bigger problem: Chinese government hackers were digging deep inside the company's systems in the United States. And the hackers weren't just going after Google's algorithms, or trying to help Baidu, the search firm China had created to compete with Google and that became the world's second-biggest search engine. The hackers were looking for intelligence—everything from the activities of Chinese nationals living in the United States to the communications of key American decision-makers who used Gmail because it was hard to access federal computers from home. The hackers mapped where they worked and what their vulnerabilities were.

The malware inserted in Google's system was encrypted and buried in corners of Google's corporate networks where it could easily be missed. Once it was dug in, the malware created a covert communications channel, or a back door, to China to siphon out whatever information the malware scooped up.

Google wasn't the only target. There were about thirty-five other companies hacked around the same time, though Google was clearly the top priority. This group of attacks was soon given the name "Operation Aurora" by Dmitri Alperovitch, then a young researcher at McAfee who years later would emerge as a key player in identifying the Russian intruders into the Democratic National Committee.

The targets of Operation Aurora pointed to China's motives. Google engineers discovered that, in addition to looking for some of the source code for Google's search engine, the hackers were trying to break into the Gmail accounts of Chinese human-rights activists as well as their supporters in the United States and Europe.

Aurora marked the first time that the Chinese were caught engineering a major hack that sought to steal information from a non-defense firm. "We never before saw a commercial company come under that level of sophisticated attack," Alperovitch said at the time. "It's a big change." In fact, this was the moment that cyberwars began focusing on what civilians kept in their networks.

"That was the surprise," one of Google's top executives said to me later. "We weren't producing the F-35. We weren't making space lasers. We weren't designing ICBMs. So it was something of a wake-up call that we were right in their gun sights."

Google took the bold step of announcing, in early 2010, that it had been targeted in a "highly sophisticated" attack that came out of China. Google notified other companies it knew had also been targeted but many of these companies did not want to be named publicly for fear of angering the Chinese or revealing their own vulnerabilities. But Adobe, whose software was crucial for making PDFs and other office documents, and a handful of other companies were willing to take the risk and name the Chinese. The companies contended that only a government actor would have the talents to conduct such a complex intrusion.

There was little doubt that the attack on Google had been ordered by the top levels of the Communist Party. A secret State Department cable in the WikiLeaks trove alleged as much, to no one's surprise: "A well-placed contact claims that the Chinese government coordinated the recent intrusions of Google systems. According to our contact, the closely-held operations were directed at the Politburo Standing Committee level."

What surprised other Silicon Valley firms doing business in China was that Google suggested it would be fighting back: it would no longer obey Chinese rules about censoring search results on Google.cn, its Chinese server. Inside Google, the chairman, Eric Schmidt, knew exactly what the company's defiance would mean. David Drummond, the company's chief legal officer, wrote in a blog: "We recognize that this may well mean having to shut down Google.cn, and potentially our offices in China."

That conclusion was likely the exact one that Li Changchun, the propaganda chief, wanted Google to reach. Since the Chinese had already replicated Google's business model with Baidu, the next step, it seemed, would be to force Google out of the market.

Schmidt told me afterward that the Aurora attacks had pretty much "ended the debate inside the company about what our future was on the mainland." If China was willing to go to all that trouble to break into the company's servers in the United States, it would clearly have no compunctions about demanding every bit of user data in China, and Google wasn't willing to let that happen. By later in 2010, Google was packing up and moving out of Beijing.

There was one more twist to the Aurora story that no Google executive revealed at the time. The Chinese had cracked a Google server that contained a database of court orders delivered to Google—the orders from the United States Foreign Intelligence Surveillance Court and other judges around the country. The FBI's counterintelligence team knew what this particular theft meant: the Chinese intelligence services were looking for evidence that their own spies in the United States had been compromised and placed under surveillance.

"Knowing that you were subjects of an investigation allows them to take steps to destroy information, get people out of the country," one former official told Ellen Nakashima of the *Washington Post*. It was a brilliant move: after years in which Chinese spies were swept up in FBI investigations, Beijing had decided to get a step ahead of the investigators. The Chinese Ministry of State Security, it turned out, was penetrating American intelligence operations—via Google.

It wouldn't be the last time.

NATURALLY, UNIT 61398—formally the 2nd Bureau of the People's Liberation Army's General Staff Department's 3rd Department—existed almost nowhere in the Chinese organizational charts. But by 2013 it had been in the sights of American intelligence agencies for several years.

The day before Barack Obama was elected president in 2008—and the same week that the Defense Department was battling the Russians—another State Department cable voiced official concerns about how frequently the unit was breaking into US government sites. Obama himself had felt the sting: the Justice Department contacted him during his 2008 campaign to explain that the Chinese were deep inside his own campaign computers, presumably looking to understand how their complex relationship with Washington would change with the election of a young senator who had barely been on China's radar.

"That was our early taste of this problem," Denis McDonough, who became Obama's chief of staff, later told me.

Everything the US government knew about the unit was still classified, owing to some strange combination of diplomatic concerns that no one would quite articulate and the fact that the United States had just launched a criminal investigation into the thefts. But in hacking investigations the government often does not have a monopoly on the evidence, since most companies not only call a private cybersecurity firm first but often hesitate to let the FBI into their computer systems for fear of what else might be exposed to federal investigators.

Which was how, by 2012, Kevin Mandia's staff had come to be looking at an actual Chinese hacker through the hacker's own camera. And the hackers had another vulnerability that helped Mandia home in on their identities. Because they had special Internet access in China, the hackers could bore through the Great Firewall and do what ordinary Chinese could not: check their own Facebook accounts. By watching the hackers type, Mandiant was able to track down their names.

Among the most colorful was a hacker whose screen name was UglyGorilla. He ranked among Unit 61398's most prolific operators and turned out all kinds of malware from an IP address that was right in the Pudong neighborhood. Mandia watched UglyGorilla and his fellow workers log in and steal blueprints and identification numbers from RSA, the American company best known for making the SecurID tokens that allow employees at military contractors and intelligence agencies to access their email and corporate networks. The hackers then used the data stolen from RSA to get into Lockheed Martin.

While Mandia was keeping an eye on UglyGorilla, another hack—perhaps the most troubling and mystifying—was taking place out of his view, in Canada. The target was a subsidiary of Telvent, a company that designs software that allows oil-and-gas operators to turn their pipelines on and off remotely and to control the flow of energy supplies. Telvent held the blueprints for half the oil and gas pipelines in the Western hemisphere. In September 2012, the company had to admit to its customers that an intruder had broken into the company's systems and taken project files.

No one could quite figure out whether this particular hack was the work of Unit 61398—which looked probable—or some other Chinese group. Nor was the motive clear. Were the hackers planning to take control of the pipelines, perhaps in time of war, and freeze out much of the United States? Or were they simply industrial thieves, looking to steal the software so that they could replicate similar pipelines in China or elsewhere? While the United States and the Canadians investigated, the findings—if any—were never made public. The mystery remains.

EVEN WHILE THE Telvent hack was under way, the Chinese government was preparing another, far more sophisticated covert operation in Washington. It would ultimately yield them a map of how the US government operates, populated with the most intimate details of the lives of 22 million Americans—almost 7 percent of the country's population.

The data were extracted from a rather boring corner of the US government, the Office of Personnel Management—a vast bureaucracy that acts as the record-keeper for the millions of people who have worked, currently work, or have applied to work for the US government, whether as employees or as contractors.

As the Obama administration turned, belatedly, to locking down the US government's cyber infrastructure after the Manning and Snowden information thefts, OPM was not exactly high on its priority list. "The first focus was on the big national security apparatus," Michael Daniel, Obama's cyber coordinator, recalled later. "Defense. The intelligence agencies. People didn't think much about OPM."

But the Chinese did. They carefully surveyed the federal government's networks and quickly figured out that buried in the aging computer systems at OPM was a huge repository of the least protected highly sensitive data collected by the US government. OPM was responsible for gathering the information needed to perform background checks on almost anyone who needed a "secret" or "top secret" security clearance. Five million Americans held those clearances in 2014, when China cracked the repository wide open.

To obtain a security clearance from the US government, prospective federal employees and contractors have to fill out an exhaustive 127-page form—Standard Form 86—in which they list every personal detail about their lives. Every bank account, every medical condition, every illegal drug they used in college. They must detail information about their spouses, their kids, their ex-spouses, and their affairs. They even have to name every foreigner they have come into close and continuing contact with for the past decade or so.

The data provided in the SF-86—and the reports of the investigators who subsequently use that information to conduct background checks—constitute a treasure trove for any foreign spy agency. Here, in one place, resides an encyclopedia of the American national-security elite: Not just names and Social Security numbers, but information about where people work, where they have been posted around the world, and whether they are so deeply in debt that they may be easy marks for recruitment. The personal histories offer a wealth of potential blackmail information, as well as clues about how to impersonate a family member or friend online.

The Chinese security services possessed a far better understanding of this vulnerability than did most members of Congress or the administration. With just a bit of exploration, the Chinese hacking team discovered that the data were being kept at the Department of the Interior—completely unencrypted—because it had spare digital storage space. That meant the records were stored in the same systems used by the national parks for tracking buffalo migration, or managing fishing stocks on federal lands.

This was the least of the problems with OPM's information-security infrastructure. The agency's IT security environment was appallingly inadequate, as the OPM's inspector general—the department's independent watchdog—had documented in a series of reports dating back to 2005. The system itself was outdated, but management made it even worse—they failed to follow nationwide government policy on security protocols, neglected to maintain their systems properly, and ignored advice on best practices. By November 2014, the problems were so acute that, in a scheduled audit, the inspector general recommended shutting down parts of the system because the holes were so big that they "could potentially have national-security implications." (And they already had; OPM just didn't know it yet.)

But shutting down the system was not an option: there was a backlog of tens of thousands of security-clearance applications. Agencies across the US government—from the Pentagon to the Drug Enforcement Agency—were clamoring to get their people cleared and pensions paid. The inspector general's shutdown recommendation was roundly rejected by Katherine Archuleta, OPM's overwhelmed director.

From the start, Archuleta and her staff were clueless about what was happening in their networks. The agency's computers had no warning system to send an alert that a foreign intruder was lurking in the system, or had begun to siphon data out of it at night. The best guess of the investigators, who later spent over a year trying to piece together a timeline of the hack—stymied not only by the limits of technology but also by a recalcitrant OPM bureaucracy—was that the hackers most likely cracked OPM's systems repeatedly in late 2013.

The Chinese got caught once and expelled, in the spring of 2014, when they got "too close to getting access" to the systems that held private personal information, a congressional investigation later discovered. But even that discovery did not lead to a crash effort to seal the system. And the Chinese now had what they needed most: a map of OPM's networks and credentials stolen from one of the agency's external contractors.

Soon the hackers were back. They logged in with the contractor's stolen passwords and dropped malware into the system to open up backdoor access. For about a year they operated undetected in the network and systematically exfiltrated the SF-86 forms and the written reports on background investigations. At some point during the summer of 2014, the SF-86 forms for 21.5 million people were copied from OPM's network. By December, 4.2 million personnel files—covering 4 million current and former federal employees, with their Social Security numbers, their medical histories, and their marital status—had been stolen. And by March 2015, 5.6 million fingerprints had been copied and spirited away. OPM itself never noticed how much data were flowing from its systems, possibly because the Chinese politely encrypted the data on the way out the door, a step that OPM itself hadn't taken to protect the mountain of sensitive information it held.

It wasn't until April 2015, when a private computer-security contractor working for OPM flagged an error on a domain name—in this case "opmsecurity.org"—that the agency's cyber team began to investigate in earnest. The domain had been operating for about a year, but no one at OPM had created it. Worse, it was registered to "Steve Rogers"—a fictional character better known for his exploits as the superhero Captain America, one of the Avengers. A second website, discovered shortly afterward, was registered to his comrade Tony Stark. Connoisseurs of hacking techniques immediately observed that a Chinese military group had, in the past, left similar odes to the Avengers.

Fifty days of radio silence followed as OPM scrambled to understand what had happened. Even other parts of the Obama administration couldn't get straight answers. The Office of Management and Budget, one senior official recalled, received conflicting information about how big the breach was. "I don't think they were lying to us," one of their senior officials said. "I think they didn't know how many computers they had, much less who was on them." The security company Cylance helped sort through the wreckage; a technician working on the case wrote a pithy email to the company's chief executive: "They are fucked btw."

That was a decent summation. But the damage wasn't limited to the employees whose data were retained by OPM. While the intelligence agencies knew better than to keep the records of their operatives on the OPM system—partly because they didn't trust it—the top two officials at the CIA, director John Brennan and the deputy director, David Cohen, quickly came to the conclusion that scores of their operatives abroad were now vulnerable. Many were posted to China under "official cover," meaning they were posing as diplomats. To make that cover convincing, they had a State Department history and a file—but sometimes with career gaps or other clues the Chinese might pick up on.

It became apparent at the CIA and other intelligence agencies that the problem was even more complex. In an age of big-data techniques, the database was far more valuable than its millions of individual files. It allowed the Chinese to compare the OPM files to their own intelligence resources and even to Facebook profiles and the digital dust that diplomats and spies left in their past postings. It was easier than ever before to unmask CIA operatives. And the problem was not limited to existing officers: those still in training, or awaiting assignments, could also be identified. Soon dozens of postings to China were canceled. As Robert Knake, a former director of cybersecurity policy issues in the Obama White House told me, "a whole bunch of CIA case officers" could be "spending the rest of their careers riding desks."

The OPM hack offered a glimpse of the future, of what happens when old-fashioned espionage meets the new world of data crunching. Investigators looked at the hack of Anthem, a health-care company, in a new light; while the OPM hack was still under way, Chinese hackers, also suspected of working for the government, had been caught after stealing upward of 78 million records. The hack raised the possibility that all of these databases were being combined to provide a deeper picture of Americans.

Adm. Michael Rogers, the NSA chief, hinted at this issue when he observed that just a decade before the OPM hack, stealing 22 million records would be of little value; whatever country got them would be overwhelmed by so much information. At a talk one evening in Aspen, Colorado, soon after the disclosures in 2015, he alluded to the larger issue, gently: "From an intelligence perspective, it gives you great insight to potentially use for counterintelligence purposes. . . . If I'm interested in trying to identify US persons who may be in my country—and I am trying to figure out why they are there: Are they just tourists? Are they there for some other alternative purpose? There are interesting insights from the data you take from OPM."

This was entirely new territory for the intelligence community—and terrifying in its scope. As word of the size of the OPM loss began to leak out, Archuleta issued ridiculous-sounding assurances, such as "Protecting our federal employee data from malicious cyber incidents is of the highest priority at OPM." History suggested otherwise. She repeatedly rejected demands from Capitol Hill that she resign. The White House declared its support for her, but she was gone by mid-July.

But, at least in public, the administration never leveled with the 22 million Americans whose data were lost—except by accident. Federal employees were sent letters telling them some of their information might have been compromised, and they were offered several years of free credit-monitoring—as if the information had been stolen by criminals. (It has never shown up on the black market, another sign the theft was an intelligence operation.) The White House refused to blame Beijing. Fortunately James Clapper, the director of national intelligence, slipped in one public interview and offered his grudging respect for tradecraft. "You have to kind of salute the Chinese for what they did," he blurted out. (He later tried to walk that comment back.)

Appearing in Congress a few weeks later, Clapper insisted that the whole incident was espionage, pure and simple, and therefore did not constitute an "attack." Over a two-hour hearing, members of Congress got angrier and angrier: it looked like an attack to their constituents, they said.

Clapper pushed back, in one of those rare moments when it became clear that the United States had no intention of agreeing to rules for behavior in cyberspace that could impede our own intelligence agencies. Having previously declared, "If we had the opportunity to do the same thing, we'd probably do it," Clapper now told the assembled senators:

"I think it's a good idea to at least think about the old saw about people who live in glass houses shouldn't throw rocks. . . ."

"So, it's okay for them to steal our secrets, that are most important," Sen. John McCain shot back, "because we live in a glass house. That is astounding."

"I didn't say it's a good thing," Clapper replied. "I'm just saying both nations engage in this."

MANDIANT AND THE *New York Times* finally published their reports on Unit 61398 in 2013, as the OPM hack was unfolding. Reading the news, David Hickton, the US attorney for the Western District of Pennsylvania, thought the biggest case his office ever handled had just been blown up.

A big, tough-talking prosecutor, Hickton was a fixture in Pittsburgh: in the mornings you could usually find him at Pamela's, a pancake place in the gritty Cemetery Hill section of town. As he read the article, with its details about UglyGorilla and his fellow hackers, "I thought, this is the end of it—we'll never catch the Chinese by surprise now."

Hickton was then in the center of a grand experiment to determine whether criminal charges could be brought against foreign governmental officials—in this case Chinese military officers—for hacking into companies in the United States. It was a case that made a lot of American officials nervous, not least at the NSA. If the United States could indict Chinese hackers for stealing intellectual property, what was to stop the Chinese from indicting members of the NSA's Tailored Access Operations unit for going inside Huawei? Or the Iranians from indicting Americans for blowing up centrifuges at Natanz?

Hickton was largely uninterested in these arguments: Political to his core, he knew what needed to be done for Pittsburgh. The city had been at the center of a number of the Chinese efforts to grab American expertise, and it was time to hit back.

There was no shortage of victims for Hickton to select from. Westinghouse, a nuclear power company headquartered in the greater Pittsburgh metropolitan area, had discovered that, while it was in the midst of building four cutting-edge nuclear power plants in China in 2010, some of its proprietary data had been stolen, including designs for the reactors. The thefts would enable Chinese competitors to acquire the same technology without spending hundreds of millions of dollars on research and development. Then, for good measure, the hackers grabbed nearly 700,000 pages of company emails, presumably looking for a glimpse of the Westinghouse leadership's negotiating strategy with a big, state-owned Chinese firm.

There were other victims: U.S. Steel, one of the few survivors from the old days in Pittsburgh, found malware on its systems while it was engaged in some unfair-trade-practices cases against Chinese steelmakers. The Chinese even stole emails from the United Steelworkers, the reeling union, about its strategies to pursue trade complaints against Chinese manufacturers.

In each case Hickton's job was to find ways to prosecute individual officers in Unit 61398 without relying on evidence from the intelligence agencies. "We needed something we could bring to court, if this ever made it to court," Hickton told me.

What would be missing, of course, were the NSA's classified intercepts of the officers inside the big white tower on Datong Road. But as Mandiant had proven, it was possible to get pictures of the perpetrators within the building—which the Chinese have yet to concede serves as the headquarters for a PLA cyber unit—without actually relying on the NSA.

Working with the targeted companies and a trail of forensic evidence, Hickton was able to identify the five PLA officers later named in an indictment, using many of the same techniques that Mandiant had used. He even had their names—Wang Dong, Sun Kailiang, Wen Xinyu, Huang Zhenyu, and Gu Chunhui—and ranks, enabling Hickton to make a public identification that, as he put it, "would freak out the Chinese." But he had no illusions about bringing the five officers to justice. Unless these five decided to take their kids to Disney World sometime in the near future, the chance of grabbing them in the United States was next to nil. The case would be more symbolic than anything else—a legal and diplomatic gamble that the publicity around the indictment, and exposure of some of the evidence, might embarrass the Chinese into halting some of their most egregious behavior.

"I can't do the diplomacy part," Hickton said. "I can only do the we're-trying-to-lock-you-up part."

At the center of the prosecution strategy was John Carlin, the head of the national-security division of the Justice Department. "We needed to push back, and to do it through our legal system," Carlin told me. "And that means building a solid case, the way you would build any other kind of solid case."

It was a sign of the sensitivity of the whole matter that Hickton spent much of his time fighting a rearguard action with the Justice Department; he didn't mind working with Carlin on the case, but the last thing he wanted to see was the case being taken away from him and enlarged in Washington. And other parts of the US government, Hickton believed, didn't want the Justice Department to dabble in state-supported foreign cyber incursions at all. "The State Department didn't like it because they were afraid it would mess up negotiations with the Chinese on other things," Hickton told me. "The intelligence guys were afraid it would shut down their sources. So I had to spend months just keeping everyone together."

When Hickton saw the Mandiant report and the *Times* investigation, it seemed to him that the work he'd done was for naught. He believed someone in Washington had made a deliberate decision "to 'out' the PLA, and that events would take their own course as a result of the report." He was wrong: there was no government leak, and as he notes, the public outing bolstered his case.

Hickton took the evidence to a grand jury. They indicted five officers of the People's Liberation Army, including "UglyGorilla" and his compatriot "KandyGoo." But the indictment was kept under seal, pending an approval in Washington that it was the right time to take on the Chinese government.

Hickton was constantly on the phone to DC, or on a plane, pushing to get the indictments announced. Finally, in May of 2014, the approval came. The big announcement, to Hickton's annoyance, came in Washington, not Pittsburgh. "State actors engaged in cyber espionage for economic advantage are not immune from the law just because they hack under the shadow of their country's flag," Carlin said in Washington. "We will hold state-sponsored cyber thieves accountable as we would any other transnational criminal organization that steals our goods and breaks our laws."

When the *Times* asked Carlin and James Comey, then FBI director, whether the Chinese might retaliate by indicting Americans who hack on behalf of the US government, they said that, naturally, they could not discuss any offensive US cyber operations. But the difference, they both stressed, was that the United States didn't steal secrets from China and give them to corporations like Google and Microsoft and Apple.

They were right, but it was a very American answer. It is a distinction that the Chinese have never bought into: To them, economic security and national security are a seamless web, and building strong, state-owned firms is essential to the defense of the state. And the indictments pointedly didn't mention Chinese attacks aimed at the Defense Department or major defense contractors: clearly the United States did not want to invite Chinese revelations about American attacks on similar military targets in Beijing, Shanghai, and Hong Kong.

UglyGorilla and his coworkers have never seen the inside of an American courtroom, and Hickton, who left his job at the end of the Obama administration, concedes they may never face a judge. But Hickton keeps a copy of one of his favorite keepsakes from the case: the big red Wanted poster the Justice Department printed with the pictures of all five PLA officers.

. . .

THE CHINESE WERE blindsided by the indictments and professed outrage, calling the specifics of the indictments "fabricated facts" that "grossly violate the basic norms governing international relations and jeopardizes China–US cooperation." It was they who were the victims of cyberattacks, the Chinese claimed, not the perpetrators. But the publication of the pictures of actual officers working at their keyboards made it clear to the PLA they were going to have to up their game. Eric Holder, then the attorney general, told me and my colleague Michael Schmidt that his response to the Chinese was in the nature of a dare: "If we fabricated all this, then come over to Pittsburgh and embarrass us by forcing us to put up or shut up, and we'll put up." The Chinese leadership didn't press the point much further—apart from the odd rhetorical posturing—but they also showed little willingness to curtail industrial espionage. For the rest of 2014 and into 2015—with the OPM revelations—the two sides settled into a tit-for-tat strategy. The only glimmer of progress came in mid-2015 when a United Nations "council of experts" began to draft rules about what kind of hacking should be off-limits. The theft of intellectual property—a violation of international law even in the pre–cyber age—was an easy one to agree on.

The stalemate was broken when American outrage over OPM ran headlong into government pageantry. Xi Jinping, settling into China's presidency, was heading to Washington in September 2015 for his first state visit—a moment of pomp and circumstance that most Americans tended to ignore but was vital to the status-conscious Chinese leadership. Chris Painter, the head of the State Department's cyber unit, recalled later that the Chinese officials were "almost pathological in wanting his trip to go perfectly."

Obama's team realized they had leverage and promptly threatened to impose sanctions on China for a variety of cyber activities, including Unit 61398's exploits, in the days just before President Xi was scheduled to arrive. They knew that to the Chinese, sanctions would cast a huge pall over the trip and would suggest that Xi wasn't in command of the relationship. The only way to avoid this embarrassment, they told the Chinese, was to negotiate the bare bones of the first arms-control accord for cyberspace.

Susan Rice, Obama's national security adviser, was dispatched to Beijing in August. All the original American intelligence assessments of how Xi would conduct himself as leader—that he would focus on domestic issues, not press for territorial gains or challenge American influence around the world—had been proven wrong. He was far more of an activist on the geopolitical stage than anyone had expected. And while Rice was escorted in for a lengthy conversation with Xi, she left with the issue of cyber espionage unresolved. It looked as if Xi was preparing to stonewall at his meeting with Obama, the last chance to get something significant going before he left office.

But when Rice returned to Washington, "they suddenly called up and said they needed to send a delegation here," she recalled later. The specter of sanctions—that would target a select set of companies and government entities that had profited from hacking American firms—just ahead of Xi's visit had finally given the Chinese pause. Suddenly fifty Communist Party officials and government bureaucrats, led by Meng Jianzhu, a close Communist Party adviser to Xi and head of state security, secretly landed in Washington to work out a deal.

Four days of marathon sessions took place at the Shoreham Hotel, near Rock Creek Park: a place so jammed with tourists of all nationalities that a delegation of that size could blend in. Painter and Suzanne Spaulding, a former senior CIA official who was overseeing cyber policy at the Department of Homeland Security, focused on a series of steps to stem the flood of attacks on American industry. "We were all thinking about OPM," Spaulding later recalled, but espionage was left off the agenda—it would have complicated an already fraught set of issues.

The talks ended at three a.m. on the morning the Chinese were scheduled to return to Beijing. Upon landing, Meng acknowledged for the first time that there was a difference between cyber espionage for national-security purposes and cyber espionage for corporate economic benefit. Obama told American business leaders that cyberattacks would "probably be one of the biggest topics," and his goal was to see "if we and the Chinese are able to coalesce around a process for negotiations" that would "bring a lot of countries along."

When President Xi himself arrived in Washington several days later for his first state visit, he was treated to a lavish state dinner. Obama had invited all the Silicon Valley royalty who were struggling in China: Mark Zuckerberg of Facebook, Tim Cook of Apple, and the chief executives of Microsoft and DreamWorks.

Before Xi left, he and Obama announced an accord that included the first curbs on using the web to steal intellectual property. Oddly, it seemed to work right away: Mandiant and other firms saw a marked drop-off in that kind of hacking by the Chinese. Painter believes that Xi looked into the future and saw that "a few years from now, people are going to be stealing industrial designs from the Chinese, and he had to get ahead of it." In fact, people have already gone after the Chinese—and most of them are Russian.

But Obama's hope of creating a model that others would follow—what Kennedy did with the Limited Test Ban Treaty more than forty years before—never materialized. The agreement with China was not expanded, and no other countries began serious discussions along similar lines.

And the other subject that Obama and Xi discussed intensely, how to manage a young, headstrong dictator in North Korea, was quickly coming off the rails as well.

CHAPTER VI

THE KIMS STRIKE BACK

AGENT LACEY (LIZZY CAPLAN): Kim Jong-un is now capable of nuking all of the West Coast. The point is we're talking about nuclear nations at war with each other . . . The CIA would love it if you two could take him out. . . .

AARON RAPAPORT (SETH ROGEN): Like, for drinks?

LACEY: No, no, no take him out.

DAVE SKYLARK (JAMES FRANCO): Take out—like to dinner?

RAPAPORT: Take him out to a meal?

LACEY: Take him out.

RAPAPORT: Like on the town?

SKYLARK: Party?

LACEY: No, uh, take him out.

RAPAPORT: You want us to assassinate the leader of North Korea?

LACEY: Yes.

SKYLARK: What?!

—From The Interview, *the 2014 comedy that prompted North Korea's cyberattack on Sony Pictures Entertainment*

MICHAEL LYNTON, the lean, European-born chief executive of Sony Pictures Entertainment, remembers well what happened when he called the State Department in the summer of 2014. He was worried about a torrent of threats from North Korea, all designed to force the studio to halt the release of a forthcoming comedy called *The Interview.*

"I had never seen a country demand that we kill a project," Lynton told me.

It wasn't hard to understand why the North Koreans were upset about the imminent release of a farce starring Seth Rogen and James Franco. The plot was not exactly subtle: Two bumbling, incompetent journalists score an interview with Kim Jong-un, but before they leave for the Hermit Kingdom they are recruited by the CIA to blow him to smithereens. The plot was completely improbable, but the North Koreans were not known for their finely honed sense of satire.

Publicity about the movie quickly pierced the cocoon of Pyongyang. The poster was arresting: Designed with Soviet-style touches from the Cold War, it depicted Kim's missiles and tanks over an image of the young North Korean leader, looking appropriately fierce. The poster turned out to be more engaging than the movie it was advertising.

North Korea's foreign ministry, anticipating the film's plot, had already written a searing letter of protest to the secretary general of the United Nations, Ban Ki-moon, demanding that he intervene to stop the movie's distribution. It apparently took awhile for the North to figure out that the secretary general, a South Korean, was not especially interested in solving their problem. And even if he had been, he was not in a position to have influence over Hollywood studios.

When the letter-writing gambit failed, North Korea began issuing threats against the United States. If Sony released the movie in American theaters as planned, on Christmas Day 2014, it would be viewed as an "act of terrorism" meriting "a decisive and merciless countermeasure." This was the kind of line the North rolled out in response to everything from military exercises to sanctions. In other words, the response sounded like a parody of the dialogue in *The Interview*.

In Washington in 2014, before Kim's missiles could credibly threaten the capital, such North Korean threats usually prompted the kind of yawns reserved for a budget hearing on agricultural subsidies. So the bluster over *The Interview* elicited no government response. But it got Lynton's attention. As a business executive and then a Hollywood studio executive, he wasn't accustomed to doing geopolitics. And the more noise the North Koreans made, the more nervous he became—in part because his bosses at the headquarters of the Tokyo-based parent company, Sony Corporation, were terrified. Its chief executive officer Kazuo Hirai was so anxious that Lynton and his co-chair, Amy Pascal, ordered the studio to tone down a scene at the end of the movie in which Kim's head appears to explode during a gruesome assassination. Soon the name "Sony Pictures" disappeared from all of the film's posters and promotional materials as the corporate leadership in Tokyo did all it could to distance the parent company from the film.

Still, the increasingly hysterical-sounding threats from North Korea left Lynton with a bigger decision—whether to kill the project altogether.

That's when Lynton called Danny Russel.

Russel was then the State Department's top Asia diplomat, a wry and experienced hand who had, by the time he turned sixty, seen just about every form of bizarre North Korean behavior. He had worked behind the scenes to get American hostages released, designed sanctions regimes, and helped draft diplomatic initiatives over the North's weapons programs that he knew the Kim family would reject. Lynton didn't know Russel—studio executives don't spend a lot of time in Foggy Bottom, and diplomats may have an understandably jaundiced view about Hollywood. But when Lynton went looking for someone in the US government to consult, everyone suggested Russel. On their first phone call, Lynton quickly got to the urgent question: Were the North Koreans simply making noise, or was the situation about to get a lot worse?

"He either asked directly or by implication if we wanted them to pull it down because of the risk of retaliation against the US," Russel recalled. Recognizing instantly that the US government couldn't get into the business of approving or disapproving movies, Russel told Lynton it was a "business decision" for Sony Pictures. "I didn't want to be in the position of having the United States government abridging free speech at the behest of a dictator," he said. "It was their decision."

Russel offered one last bit of advice: Don't take publicity photos of Rogen and Franco at the Demilitarized Zone. The North Koreans get a little touchy up there. But as the phone call ended, Russel shared with Lynton the Washington wisdom on North Korea's hyperbolic warnings: Most of them, he said, "were bullshit."

What neither Russel nor Lynton knew was that North Korea's small army of hackers had already begun figuring out how to decimate Sony. "At that point in time, Kim Jong-un was relatively new in the job, and I don't think it was clear yet how he was different from his father," Lynton said. "Nobody ever mentioned anything about their cyber capabilities."

NOBODY MENTIONED NORTH KOREA's cyber skills because no one was really paying attention. And by the time *The Interview* was being made, the Hermit Kingdom had gone from viewing the Internet as a threat to viewing it as a brilliant invention for leveling the playing field with the West.

Like the Chinese, Kim Jong-il, the son of the country's founder and the father of its current leader, had initially seen the Internet as a threat to his regime; anything that allowed citizen-to-citizen communication could complicate ironclad control over the country. In North Korea, unlike in China, the Internet was not difficult to rein in, at least before smartphones began slipping over the Chinese border. North Korean households had no computers, just televisions and radios with a couple of state-run channels.

But over time, even a sealed-off regime began to see the merits of using the Internet to wreak havoc and make profits. Kim Heung-kwang, a North Korean defector who said in an interview with the *Times* that he helped train many of the North's first cyber spies, recalled that in the early 1990s a group of North Korean computer experts came back from China with a "very strange new idea": using the Internet to steal secrets and attack the government's enemies.

"The Chinese are already doing it," he remembered one of the experts saying.

The North Korean military began training computer "warriors" in earnest in 1996, he recalled, and two years later opened Bureau 121, now the primary cyberattack unit. Members were dispatched for two years of training in China and Russia. Jang Sae-yul, a former North Korean army programmer who defected in 2007, said these prototypical hackers were envied, in part because of their freedom to travel.

"They used to come back with exotic foreign clothes and expensive electronics like rice cookers and cameras," he said. His friends told him that Bureau 121 was divided into different groups, each targeting a specific country or region, with a special focus on the United States, South Korea, and the North's lone ally, China.

"They spend those two years not attacking, but just learning about their target country's Internet," said Jang, who was a first lieutenant in a different army unit that wrote software for war-game simulations. As time went on, Jang said, the North began diverting high school students with the best math skills into a handful of top universities, including a military school that specialized in computer-based warfare called Mirim University, which he attended as a young army officer. Others were deployed to an "attack base" in the northeastern Chinese city of Shenyang, where there are many North Korean–run hotels and restaurants.

Before long Kim Jong-il himself started sounding like a cable-television pundit on the subject of cyberattacks: "If warfare was about bullets and oil until now," Kim allegedly told top commanders in 2003, according to Kim Heung-kwang, "warfare in the twenty-first century is about information."

It's unclear whether Kim Jong-il ever really believed his own bromides about information warfare—or had much of an idea how to turn the slogan into a strategy. At the end of the day, he relied on his nuclear arsenal to keep his regime in power and his family alive. But he was convinced that it was worth identifying promising students at an early age for special training in the hacking arts. Their first step was China's top computer science programs.

Then the FBI's counterintelligence division noticed that North Koreans assigned to work at the United Nations were also quietly enrolling in university computer programming courses in New York. James Lewis recalled that as the enrollment numbers for the North Koreans rose, "the FBI called me and said, 'What should we do?'"

"I told them, 'Don't do anything. Follow them and see what they are up to.'"

What they were up to didn't seem very scary at the time. But North Korean engineers learn fast—ask any missile scientist—and they got good quickly. "There was an enormous growth in capability from 2009 or so, when they were a joke," said Ben Buchanan, a researcher at the Cyber Security Project at Harvard who has written extensively on the dilemmas of protecting networks in a world of cyber conflict. "They would execute a very basic attack against a minor web page put up by the White House or an American intelligence agency, and then their sympathizers would claim they'd hacked the US government. But since then, their hackers have gotten a lot better."

No one in Washington seemed especially alarmed. A National Intelligence Estimate in 2009 wrote off the North's hacking prowess, much as it underestimated the speed at which the country's long-range missile program would come to fruition. It would be years before the North Koreans could mount a meaningful cyber threat, it concluded.

The assessment might have been accurate—had Kim Jong-il lived. When Kim Jong-un succeeded his father in 2011, few expected that an inexperienced, narcissistic twenty-seven-year-old who had not been groomed for the job could establish his authority with the North Korean military and the country's elite. He surprised everyone. His first task, he knew, would be to make North Korea's nuclear threat a credible one. His second was to eliminate potential rivals, which he sometimes did with an antiaircraft gun. His third was to build a cyber force, and he brought to this task a sense of urgency.

By the time Kim Jong-un came to power, Bureau 121 had been up and running for more than a decade. And while Kim is often caricatured as a buffoon in American pop culture, he deftly seized on an asymmetric capability that his father and grandfather—the Dear Leader and the Great Leader, respectively—had never exploited. At Kim's direction, the North built up an army of upward of six thousand hackers, mostly based outside the country. (They eventually spread from China to the Philippines, Malaysia, and Thailand, all countries that advertise something in short supply in North Korea: beach resorts.)

The idea was to make cyber offense more than just a potential wartime weapon; like the Russians, Kim saw the opportunity for theft, harassment, and political score-settling. "Cyberwarfare, along with nuclear weapons and missiles, is an 'all-purpose sword' that guarantees our military's capability to strike relentlessly," he reportedly declared, in comments that were later relayed by a South Korean intelligence chief.*

By 2012, Kim had begun dispersing his hacking teams abroad. China was the first stop, the closest country with an Internet infrastructure that could sustain substantial malicious activity while giving the North's hacking teams plausible deniability. But over time the hackers spread to India, Malaysia, Nepal, Kenya, Poland, Mozambique, and Indonesia—places that often took North Korean laborers. In some countries, like New Zealand, North Korean hackers were simply routing their attacks through the country's computers from abroad. In others, like India, where nearly one-fifth of Pyongyang's cyberattacks now originate, the hackers were physically stationed in the country.

Success seemed easy and cheap. "You could argue that they have one of the most successful cyber programs on the planet, not because it's technically sophisticated, but because it has achieved all of their aims at very low cost," said Chris Inglis, a former deputy director at the National Security Agency.

To some degree the North Koreans learned from the Iranians, with

---

* It is almost impossible to verify what Kim actually said, unless his comments are made in statements broadcast by KCNA, the North Korean news agency.

whom they have long shared both missile technology and a belief that the United States is the source of their problems. In the cyber realm, the Iranians taught the North Koreans something important: When confronting an enemy that has Internet-connected banks, trading systems, oil and water pipelines, dams, hospitals, and entire cities, the opportunities to cause trouble are endless.

North Korea's first big strike came in March 2013, seven months after Iran's attack on Saudi Aramco. During joint military exercises between American and South Korean forces, North Korean hackers, operating from computers inside China, deployed a cyberweapon very similar to Iran's against computer networks at three major South Korean banks and South Korea's two largest broadcasters. Like the Saudi attacks, the North Korean attacks on South Korean targets—an operation quickly dubbed "Dark Seoul"—used wiping malware to eradicate data and paralyze business operations. It may have been a copycat operation, but it was an impressive one. Robert Hannigan, who later tracked the North Koreans as the head of GCHQ, the British equivalent of the NSA, saw a parallel that was too dramatic to ignore.

"We have to assume they are getting help from the Iranians," Hannigan concluded.*

"It crept up on us," he said of the North Korean threat. "Because they are such a mix of the weird and absurd and medieval and highly sophisticated, people didn't take it seriously. How can such an isolated, backward country have this capability? Well, how can such an isolated, backward country have this nuclear ability?"

FORTUNATELY, THERE WERE chinks in North Korea's system. And at Fort Meade, where the NSA and Cyber Command worked side by side, the urgency was building to exploit them with the same gusto that

---

* So far there has been no evidence of such help, but the hacker community is a pretty fluid one. It is just as likely the North Koreans got help from the Chinese, Russians, and Eastern Europeans.

had motivated Operation Olympic Games. The US sought to throw a wrench into North Korea's development of a nuclear capability that could be demonstrated to reach the United States.

But in the case of North Korea, the problem was far more complex than anything the Tailored Access Operations unit, or the early iterations of US Cyber Command, had faced in Iran. By mid-2013, it became clear to the Obama administration that it was simply too late to stop North Korea's bomb production. The Kim family was way, way ahead of the mullahs. While the Iranians were still struggling to make centrifuges spin to produce uranium, the North Koreans were churning out atomic bombs. Though intelligence estimates differed, the North Koreans already possessed upward of a dozen nuclear weapons—and production was speeding up.

So the focus of America's cyber warriors, as former secretary of defense William Perry put it, "had to be on the missiles that could get to the United States, because that's the only thing left for the North Koreans to complete."

For Kim Jong-un, the ability to reach an American city with a nuclear warhead was all about survival—but it was also about future power. He accelerated the effort drastically, turning it into the North Korean version of the Manhattan Project. That meant putting equal effort into a missile program that could get the weapons to the other side of the Pacific. And by 2013, for the first time, the missile program looked genuinely threatening.

"I heard Obama say more than once that he would have no problem decapitating Kim Jong-un's leadership circle if he had the chance—and thought it wouldn't start a war," one of Obama's aides later recounted. No one could provide that assurance, and because Obama was cautious to a fault even some of his own aides wonder whether he would have pulled the trigger. But he was certainly willing to do what he could to slow the North's nuclear program.

Which is why the president who had seen the power of America's newest weapon during Olympic Games, suddenly began demanding a way to down North Korea's missiles without firing a shot.

"By the end of 2013 we knew we had to do something," one senior aide said. The defense secretary, Ashton Carter, began calling meetings focused on one question: Could a crash program slow the North's march to obtaining an intercontinental ballistic missile?

Early in 2014, Obama presided over a series of meetings to explore a range of options. The Pentagon and American intelligence agencies, he decided, should step up a series of cyber- and electronic-warfare strikes on Kim's missiles, starting with an intermediate-range missile called the Musudan. The hope was to sabotage them before they got into the air, or to send them off-course moments after launch. The further hope was that the North Koreans, like the Iranians before them, would blame themselves for manufacturing errors.

It would take a year or two, Obama was warned, before anyone would know if the accelerated program could work. Only in retrospect is it clear that in 2014 Obama and Kim were using cyberweapons to go after each other. Obama's target was North Korea's missiles; Kim's was a movie studio intent on humiliating him. Eventually, each would begin to discover what the other was plotting.

BY MID-2014, NORTH Korea had picked its next target for a cyberattack—and it was in London.

As the North Korean protests over *The Interview* were escalating in the summer, one of Britain's commercial television networks, Channel 4, announced plans to broadcast a juicy series about an American president and a British prime minister who joined forces to free a nuclear scientist kidnapped in Pyongyang.

Just as it had done with the United Nations, the North sent a letter to 10 Downing Street and demanded that the British prime minister shut down production and punish the producers. The series would be "a scandalous farce," they said. Naturally, the British responded the same way the UN did—with silence.

Within weeks, things started going wrong at Channel 4. It became clear that someone had hacked into the channel's computer systems, though the attack was stopped before it inflicted any damage. The chief executive of Channel 4 said he would not be deterred, and the production would go forward. (It didn't: months later the project was canceled, largely because the financing dried up. Fears of North Korea's reaction appeared to be among the many reasons.)

It wasn't until years later that anyone noticed the similarities between the attacks on Channel 4 and what was happening more than five thousand miles away, on Sony's storied back lot.

By the end of the summer, the hackers were boring into Sony's systems and preparing their attack. But the Obama administration was focused on another North Korea drama, one that seemed a lot more familiar. For more than a year, the White House had tried to negotiate the release of two Americans who had been imprisoned by North Korean authorities. Obama decided to send James Clapper, the director of national intelligence. Clapper was a complete anomaly in the Obama administration: He was old enough to be the father of many administration officials—the president included—and the grandfather of many of the staff. He was bald, gruff, taciturn, and the product of years in the air force—where he had served as a lieutenant general, living all over the world, including in South Korea for a stint in the late '80s. Clapper had been in North Korean territory only once before— illegally, in 1985, when he was in a military helicopter that veered into North Korean airspace.

This time, he had been invited. So in early November 2014 he landed in Pyongyang in a US government aircraft and was whisked to a state guesthouse on the edges of the capital. In the car along the way the minister of state security began peppering Clapper with questions about whether he had arrived with a major diplomatic offer in hand. "They were expecting some big breakthrough," Clapper later recalled with some wonderment. "I was going to offer some big deal, I don't know, a recognition, a peace treaty, whatever. Of course, I wasn't there to do that, so they were disappointed."

On his first evening, Clapper found himself having dinner with his

North Korean counterpart, the chief of the Reconnaissance General Bureau, Kim Yong-chol.

The traditional Korean meal was spectacularly delicious, Clapper later said, basing his culinary assessment on his years in the South. But that turned out to be the best part of what became a highly unpleasant evening. General Kim, a member of the leadership's inner circle, "spent most of the meal berating me about American aggression and what terrible people we were." He told Clapper that Washington was constantly plotting to overthrow the North Korean regime, a charge that is not entirely without merit.

Clapper shot back that it was time for the North to stop starving its people, building gulags, and threatening nuclear holocaust. From there, the discussion went downhill.

"These were the real hardliners," Clapper concluded.

But at no point in their hours of conversation did *The Interview,* or the North Korean threats against Sony, much less the North's breach of the company's systems, come up.

"I had no idea what was happening back at Sony," he told me later. "Why would I?"

He was right: that was not how US surveillance systems were set up. The United States had spent more than six decades deploying a vast surveillance capability against North Korea—the NSA was created in the midst of the Korean War—but was almost entirely focused on traditional threats. Hackers working from laptops somewhere in Asia were not the kind of security threat this apparatus was established to detect. And movie studios weren't the targets the American intelligence community was focused on protecting. In fact, because the law prevents the NSA from conducting surveillance on American soil, it could not look into Sony's networks.

The day after the dinner, Clapper won the release of the Americans and loaded them onto his plane. But before he left, he had one more encounter with the North Koreans. Along with the newly released Americans, North Korean officials handed Clapper a bill—for his share of dinner with the head of the Reconnaissance General Bureau, along with his room in the state guesthouse and the parking of his aircraft.

"I had to pay in greenbacks," Clapper later told me. "And it wasn't a small amount."

CLAPPER'S HOST, GENERAL KIM, likely knew a lot about the Sony hack well before he invited his American visitor to dinner: American intelligence officials now believe that the hackers were working, directly or indirectly, for the Reconnaissance General Bureau. But at the time, the North seemed like America's least likely concern in cyberspace. After all, who frets about cyberattacks from a country with fewer IP addresses than most city blocks of New York or Boston?

In retrospect, there was a lot that American officials should have been worrying about. Kim might be broke, and living in his own bubble of national adoration, but in 2014 he clearly understood the new contours of national power. He had correctly calculated that a cyber arsenal was the great leveler: It was dirt-cheap. He could launch it from outside the country. And unlike his nuclear arsenal, cyberweapons could be used against his greatest enemy—the United States—without fearing that fifty minutes later his country would be a smoking, radioactive cinder just north of Seoul. Kim recognized that the inevitable US threats of imposing additional economic sanctions against the North for malicious cyber activity were largely empty.* In short, cyberweapons were tailor-made for North Korea's situation in the world: so isolated it had little to lose, so short of fuel it had no other way to sustain a conflict with greater powers, and so backward that its infrastructure was largely invulnerable to crippling counterattacks.

---

*   He was right: In late 2017—after blaming North Korea for a global cyberattack called "WannaCry"—Thomas Bossert, the Trump administration's homeland security adviser, justified the United States' failure to retaliate by admitting there was little way to strike back. "President Trump has used just about every lever you can use, short of starving the people of North Korea to death, to change their behavior."

Even Kim's growing cyber army was a recognition that the United States and its allies would probably spend the next few years debating how to strike back for an attack that doesn't leave visible, smoldering ruins.

And even if the United States was willing to retaliate, Kim calculated, doing so would not be easy. To most of the world, the absence of computer networks, of a wired society, is a sign of backwardness and weakness. But to Kim, this absence created a home-field advantage. A country cut off from the world, with few computer networks, is a lousy target: there are simply not enough "attack surfaces," the entrypoints for inserting malicious code, to make a retaliatory cyberattack on North Korea viable.

Or, as one senior official of the US Cyber Command put it at dinner in Washington one night, "How do you turn out the lights in a country that doesn't have enough power to turn them on?"

Four years before the Sony hack, the United States had tried to answer that question in a secret operation code-named "Nighttrain." The agency painstakingly drilled into the networks that connect North Korea to the outside world, mostly through China. They tracked down North Korean hackers, some of whom worked from Malaysia and Thailand, hoping to identify and locate members of the North Korean cyber army. And, without telling the South Korean government—its critical ally in taking on the North Korean challenge—the US piggybacked on a South Korean cyber intrusion into North Korea's intelligence networks.

The purpose of Nighttrain—an operation revealed only in a brief and partial glimpse provided by a few documents in the Snowden trove—remains unclear. As does the NSA's motivation in going behind South Korea's back: why hadn't the United States trusted its South Korean allies enough to join forces with them, openly, in piercing North Korea's networks? Presumably the operation was intended to glean what it could about the North's leadership, about its newly formed cyber corps, and of course about its nuclear secrets. It is not clear what the United States actually gained from the effort. But whatever its success or failure, Nighttrain yielded no advance warning of what was about to happen to Sony.

ONCE THE NORTH KOREAN hackers had burrowed into the Sony networks in the fall of 2014, they littered the company with "phishing" emails, betting that someone at the studio would click on the bait. It didn't take long for that tactic to pay off. And once inside the system, Kim's hackers obtained administrator privileges, the ability to roam throughout the system. Over the next few weeks, Sony's unseen invaders mapped out where people stored emails, how the systems worked, and where Sony locked up forthcoming movies. Soon, the hackers owned the studio's system.

The North Koreans had been enormously patient. They waited until the right moment to execute each step of the attack. This was a sign of true professionals. As one senior American intelligence official put it to me: "You don't freelance in the North Korean system." But even as the North's hackers slowly broke down the last remaining barriers in Sony's network and began to crawl around, mapping their attacks, no one in the company noticed. They were the digital equivalents of cat burglars who were extremely careful not to set off a hidden alarm.

Remarkable as it seems today, no one at Sony was even thinking about its computer networks as a vulnerability. It was a stupid mistake, but hardly an uncommon one—many other companies, and the US government, would make it time and again over the next three years.

DAYS BEFORE THE 2014 Thanksgiving holiday, Lynton was driving to work in his Volkswagen GTI when he got a phone call from the office. It was David Hendler, the studio's chief financial officer. There had been a cyber intrusion into the studio's central computers. No one was quite sure of its scope, Hendler said, but it looked unusual. Maybe, he said, it would prove to be a passing issue that the IT guys could clean up by lunch. But it didn't look that way; in fact, the studio's IT department was preparing to take all of Sony Pictures offline to prevent the damage from getting worse.

By the time Lynton arrived at his office in Culver City's Art Deco Thalberg Building—on the studio lot where Louis B. Mayer once ruled Hollywood—there was no illusion that this attack would be over by lunch. "Clearly, no one had a handle on the scope of the thing," Lynton told me. And it didn't look like your ordinary cyberattack. For one thing, across Sony's campus, thousands of the company's computer screens were showing an image of Lynton's head, grotesquely severed.

Disgusting as it was, the picture was merely a distraction. While the computer users were trying to figure out what was happening, their hard drives were spinning away, wiping out whatever data were stored on them. The only employees who saved their information were those with the presence of mind to reach behind their computers and unplug their machines, bringing the hard disks to a halt. Those who stopped and stared at the image lost everything.

Shortly after Lynton arrived, Sony disconnected all of its computer systems around the world. No email. No production systems. No voicemail.

Lynton prided himself on his coolness under pressure; this wasn't his first corporate crisis. His instinct, typical of most corporate executives, was to keep the problem locked up inside Sony. After all, it would only feed the hackers' egos to figure out how much damage they had done. But the world would know soon enough.

Lynton alerted the FBI, and they set up camp on the studio lot, where many a movie depicting G-men pursuing bad guys had been filmed over the decades. But no one was spending much time ruminating on the life-imitates-art ironies. Instead, the agents quickly became preoccupied with a group that called itself the "Guardians of Peace," who began leaking out Sony's emails a few at a time. Clearly these had been swept up in the hack. And clearly someone involved in their release understood what would prove to be catnip for supermarket tabloids—indicating that whoever was behind the operation had some America-savvy help.

The Sony experience wasn't a first: the WikiLeaks publication of State and Defense Department cables in 2010 proved how easy it was to grab headlines with confidential communications stolen from a computer system. Over the following few weeks, the Sony hackers doled out emails with embarrassing details about the studio, along with Sony contracts, a few medical records, and plenty of Social Security numbers. The North Koreans had even grabbed five yet-to-be-released movies, including *Annie*.

The Sony emails attracted an audience that no State Department cables could ever match. After all, the juiciest of State Department cables dealt with complaints about embassy amenities, backbiting office politics in Foggy Bottom. The Sony emails, in contrast, included one from a studio executive who described Angelina Jolie as a "minimally talented spoiled brat." There was information about salaries at the studio and gossip about the offscreen affairs of actors and producers. There was even a leaked email that appeared to have been from the hackers themselves, sent on November 21, warning that if the studio didn't pay an unspecified ransom, "Sony Pictures will be bombarded as a whole." It looked like the email was never read before the attacks. And then there was the most ironic of all: emails from Seth Rogen to studio executive Amy Pascal, complaining about the changes Sony was making to the script of *The Interview*. "This is now a story of Americans changing their movie to make North Koreans happy," he complained.

The salacious stuff, of course, overwhelmed most coverage of the destructive power of the attack. In the space of just a few months, North Korea—a country that could barely feed its people—had struck an iconic American studio with the most sophisticated cyberattack since Olympic Games. Sony had been asleep at the wheel. As had the US government.

In retrospect, the Sony hack was a harbinger—a destructive attack that melted physical equipment utilizing only ones and zeros, as Stuxnet had done; a distracting release of private communications that dominated the news and upended careers; and a ransom demand that distracted from the real purpose of the operation. But no one knew all of this at the time. When it happened, the attack seemed like a bolt from the blue, a wild overreaction to a Hollywood comedy from the touchy thirtysomething leader of a paranoid, starving nation that was wasting its hard currency, and its scarce talents, building nukes and missiles.

With 70 percent of its computer power paralyzed, Sony had to search the world for new equipment. Meanwhile, the accounting department decided to dig through the basement to look for the old machines they once used to issue paychecks. Clearly, they wouldn't be making electronic transfers for a while.

AT THE WHITE HOUSE, the Sony attack raised a series of uncomfortable questions that would bedevil the US government many times over the next few years. From the moment the FBI camped out in the studio lot, the prime suspect was North Korea. But as President Obama's aides knew, suspecting was one thing. Proving it was another. And even if Adm. Rogers, the director of the NSA, walked into the Situation Room with incontrovertible evidence, how could it be made public without revealing the agency's sources? And what could you do to retaliate?

"It's a classic problem," Michael Daniel, Obama's cybersecurity czar, said to me during the Sony investigation. "As soon as you declare who was behind a cyberattack, the next question is always: how are you going to make them pay? And there's not always an easy answer."

In fact, the NSA was already looking back at the reams of data collected from a series of intelligence operations inside the North's computer networks, including Nighttrain, in an effort to make a conclusive case that North Korea's leadership had ordered the Sony attack. Before long they discovered that some of the tools used against Sony had been used in previous attacks mounted by North Korean hackers.

"We found what we were looking for very quickly," one White

House official told me, describing evidence that appeared to provide a direct communications link between the Reconnaissance General Bureau and the actual hackers. Even today, the US government has never revealed its evidence—in the Sony case or in other instances of North Korean hacking—because it does not want to tip its hand about what kind of monitoring may be ongoing. But it seems clear the United States uncovered some intercepted voice communications or written instructions straight from the North Korean leadership.

The evidence was persuasive enough that President Obama was briefed on it almost immediately.

"I never thought I'd be here briefing on a bad Seth Rogen movie, sir," one of Obama's aides told him as the plot became clear.

"How do you know it's a bad movie?" Obama asked.

"Sir, it's a Seth Rogen movie. . . ." Laughter broke out in the Oval Office.

But the proof only complicated the debate. The United States had lots of plans for how to respond to an attack on critical infrastructure, from dams to utilities. Clearly, the Sony attack was not in that category.

"This was a destructive attack," said Robert Litt, the general counsel for the director of national intelligence. "But you couldn't argue that it hit a vital sector of the US infrastructure. It wasn't exactly taking out all the power from Boston to Washington. So the issue was: is this the government's responsibility to defend?"

That was only one of the questions hanging in the air as Obama and his aides descended into the Situation Room on December 18. In a spirited debate, some of Obama's aides argued that whether or not the target was "critical," the United States had just been attacked.

"I remember sitting there while some of our colleagues argued that this was just like planting a bomb inside Sony, which we definitely would have categorized as terrorism," said one national-security aide who was sitting along the back bench as the argument raged. "But in this case, there was no explosion—just people operating by remote control to accomplish the same result."

Ever cautious, Obama came to the conclusion that it wasn't terrorism; it was more like "cyber vandalism," as he said a few days later. (He soon came to regret the line.) Obama did not want to escalate. But he also did not want to go through another country's networks to get inside North Korea.

"The problem," one participant in the meeting later told me, "was that the only way to go into the North Korean networks was through China, and no one wanted to have the Chinese thinking that we were attacking them or using their networks to attack someone else."

But Obama was animated by one aspect of the attack. What made the Sony strike different, in his mind, was that it was intended as a weapon of political coercion. The constitutional lawyer in him was determined not to let a dictator in a faraway, broken nation kill a movie he found politically objectionable.

Meanwhile, the threats had grown more violent. The Guardians of Peace issued a declaration that the movie's premiere in New York could be the target of a terror attack: "Soon all the world will see what an awful movie Sony Pictures Entertainment has made," the statement said. "The world will be full of fear. Remember the 11th of September 2001."

The 9/11 reference immediately heightened the stakes. Lynton suspended the release of the movie. And Obama, meeting in the Situation Room the day the threat was delivered, realized he could no longer remain silent. If he ignored a crude threat of terrorist action against theaters, he would look weak. He needed to call out the North Korean leadership, blame them for the cyberattack, and make clear what would happen if theaters were attacked. That meant he had to make it clear that the United States had linked the attack and the threats to Kim Jong-un.

But his intelligence officials were adamant that he could not reveal the presence of any implants the United States or South Korea had lurking in North Korea's systems. In fact, they did not want him to explain in public even the obvious stuff: how they had matched the hacking tools used against Sony to others previously utilized by the North Koreans, specifically by Bureau 121, which ran the country's army of cyber warriors—though the US didn't have enough evidence to attribute the Sony attack to Bureau 121 with certainty.

"It was a classic debate," one participant later told me. "The intel guys didn't want to say anything—they are wired that way. The political and strategic types wanted to create some cost for the North Koreans." But the options presented by Rogers—a counter cyberstrike on the North, or going after Kim Jong-un's accounts around the world—were difficult to accomplish and seemed likely to impinge on Chinese sovereignty. So Obama decided to name and shame the North Koreans and figure out the penalty later.

The next day, December 19, hours before leaving for vacation in Hawaii, Obama stepped into the press room and took the unprecedented step of blaming North Korea for the attack. He vowed that a proportional response would happen "in a place and time and manner that we choose." Some elements of that punishment would be visible, he said, and some would not be. He used the language of military retaliation, but without the real threat of action.

"We cannot have a society in which some dictator someplace can start imposing censorship here in the US," he said, leaving no doubt he was directly challenging Kim Jong-un. He also took a shot at Sony for pulling the film out of theaters. American filmmakers and distributors, he said, should not "get into a pattern where you're intimidated by these kind of criminal attacks."

Lynton was flummoxed by Obama's comments; he thought he was being cautious, protecting theatergoers, and he had already vowed to distribute the film one way or another. "I certainly am not planning on caving to the North Koreans," he told me.

Lynton sent his staff scrambling to find independent theaters to show *The Interview*. More important, he twisted arms to get a digital release of the movie at the same time it was appearing in the theaters. At the time, that was still extraordinarily rare in the movie business.

But this was an extraordinary circumstance. While some online movie distributors balked, Google came through, as did YouTube. On Christmas Day, after opening stockings and gifts, Americans downloaded the movie in living rooms across the country. It was still a ridiculous plot. But at least Kim Jong-un didn't win. For now.

THE SONY ATTACK was hardly the only short-of-war assault on American targets in Obama's second term, and it certainly wouldn't be the last. Neither was it perfect. Jim Lewis later concluded that the North Koreans had a few weaknesses of their own. Specifically, they thought they were stealthier than they really were.

North Koreans had not expected the United States to conclude so quickly that Pyongyang was behind the attack, Lewis told me. "Sony shocked [North Korea] when it discovered that they were not invisible in cyberspace," he wrote. But inside the White House and the NSA, the attack illuminated weaknesses in American defenses that would only grow more glaring.

The first was a deep confusion—in both the government and the corporate world—about who is responsible for defending against attacks on corporate America.

The issue had come up repeatedly. When the Iranians froze the banking networks of Bank of America and JPMorgan Chase, Obama and his aides were concerned, but they concluded that the denial-of-service attacks didn't rise to the level of requiring a national response. The attacks were viewed as crimes, not terrorism, and referred to the Justice Department, which ultimately indicted Iranian hackers.

But in early 2014, when the Iranians melted down computer equipment at the Sands Casino in Las Vegas, the administration again did not respond—even though that attack was more damaging and an act of political retaliation. The Iranians attacked the casino to show owner Sheldon Adelson that if he wanted to advocate setting off a nuclear weapon in the Iranian desert, he had better be prepared to see his prize casino go offline. That, too, was treated as a criminal act—to be dealt with in the courts—rather than as an attack on the United States.

In short, until the Sony attack Obama believed corporate America should take responsibility for defending its own networks, just as they take responsibility for locking their office doors at night. That approach made sense most of the time: Washington could not go to DEFCON 4 every time someone—even a state—went after part of the private sector. Clearly, the government could not protect against every cyberattack, just as it could not protect against every car theft or house burglary.

But the government is, of course, expected to protect against—or at least respond to—armed attacks on American cities. So what was a cyberattack more like? A home burglary or a missile attack from abroad? Or was it something completely different? And when was the potential peril to the United States so great that the government could no longer rely on companies or individual citizens to defend themselves but had to respond?

In eight years in the White House—years in which cyber went from a nuisance to a mortal threat—neither Obama nor the bureaucracy ever formulated a satisfying answer to those questions. Clearly, the first line of defense had to be the companies themselves. It made sense that when banks came under denial-of-service attacks, or utilities saw malware being implanted in their systems, the United States should hold back. After all, if corporate America thought the government was going to deal with cyber threats, they wouldn't invest in protecting themselves. And many companies didn't want the government inside their systems, even to play defense.

But when, exactly, would the United States intervene? Run-of-the-mill DoS attacks are one thing; attacks that threaten to turn off the power or freeze the financial markets are another. In the Sony case, Obama's answer seemed to be that the United States would get involved when a fundamental American value—in this case, freedom of speech and assembly—seemed threatened by a foreign power. But he never publicaly justified that conclusion—or explained which other attacks might rise to the level of a federal response. Understandably, the government did not want to draw bright lines, for fear that attackers would walk right up to them. But Americans need to know who is responsible for protecting us and our data—just as we need to know that it is the police who protect us against home invasions and the Pentagon that defends us against intercontinental ballistic missiles.

The closest the government came to drawing clear lines came when Ashton Carter, the secretary of defense in Obama's last years in office, presented a new cyber strategy at Stanford in April 2015—a strategy in which the military would take a larger role in defending American networks. "The cyber threat against US interests is increasing in severity and sophistication," he told a Silicon Valley audience that was clearly divided on the question of how much it wanted the Pentagon to be involved in policing US networks. "While the North Korean cyberattack on Sony was the most destructive on a US entity so far, this threat affects us all. . . . Just as Russia and China have advanced cyber capabilities and strategies ranging from stealthy network penetration to intellectual property theft, criminal and terrorist networks are also increasing their cyber operations."

He went on to describe the "Cyber Missions Forces" that were being built up at US Cyber Command: 6,200 American warriors, including defensive teams and combat teams. And he called on Silicon Valley to send some of its best and brightest for tours of a year in the Pentagon so that the United States would be on the cutting edge of defense. (Silicon Valley executives were cautious about this idea. The head of a major firm doing business with the Pentagon told Carter's entourage that he doubted his talented group of coders would be able to obtain security clearances. "Because they smoked pot in college?" he was asked. No, the executive clarified, "because they smoke pot on the job, while programming.")

The new Pentagon policy described by Carter left deliberately vague when those teams of cyberwarriors would be called into action. Some of Carter's aides spoke about employing American government defenses to protect against the top 2 percent of attacks—those that threaten America's vital national interests. That made sense, though that likely meant the government would step in to defend the country only for a much smaller fraction of cyberattacks, maybe the top .02 percent. And what of the other 99.98 percent? Could companies be expected to defend themselves? And what would that defense look like?

The policy revived a long-simmering debate about the advisability of letting companies go beyond building bigger defenses to actually striking back themselves at their attackers—something called "hacking back." It's illegal, just as it's illegal to break into the house of someone who robbed your house in order to retrieve your own property. But the fact of its illegality didn't stop more than a few companies from trying, often through offshore subsidiaries or through proxies. (Google engineers thought seriously in 2009 about doing harm to servers in China where attacks on the company had originated, before cooler heads prevailed.) Periodically there have been movements in Congress to make hacking back legal—often under the rubric of "active defense"—as a way of letting cyber victims create some deterrence. Regardless of whether it would work or not, hacking back would certainly be satisfying for companies. It could also start a war.

"It would be a total disaster," one senior military strategist said to me when the issue came up anew around the JPMorgan and Sony attacks. "Imagine a company takes out a big server in Russia or North Korea," the official went on. "The Russians or the Koreans see it as a state-sponsored attack. So they escalate . . ." Before the first meeting on the confrontation is held in the Situation Room, a full-scale conflict ensues, all for a retaliatory strike the president was never so much as consulted about.

That prospect—how a cyberwar could turn into a shooting war— leaves a lot of people very scared. "We need arms control," Brad Smith, the general counsel of Microsoft, told me in the weeks after the Sony hack.

But just as America did not want to discuss limiting nuclear weapons when it thought it was leading the world in the 1950s, it does not seem interested in any agreement that would limit its ability to develop its cyber arsenal. Instead, it was trying to develop new tools, and re-purpose old ones, to retaliate—deterrence by the threat of force. The possibility of arms control appeared to be off the table.

ALMOST AS QUICKLY as Obama stepped into *Air Force One* for his Christmas vacation, the blogs and Twitter feeds erupted with doubters. Wasn't this just the cyber equivalent to the faulty case against Iraq in 2003—when America was also told that the evidence against the Iraqi regime was too sensitive to reveal? Since hacks are notoriously hard to trace, how come the president was so sure about North Korea? Responses such as these revealed how deeply entrenched distrust of America's intelligence agencies had become in the post-Snowden era.

Obama had made a critical mistake: he had accused an adversary of attacking the United States, but he omitted the evidence.

The White House was caught flat-footed. It hadn't imagined, even in a post-Iraq age, that there would be serious doubts about the president's accusation. But, of course, there were many doubts. A smattering of cybersecurity firms and private investigators came out with alternative theories. Some said it was the Chinese. Or the Russians. Or a disgruntled insider. Even *Wired* magazine, usually pretty careful about such topics, characterized the case as "flimsy."

The truth was that the Obama administration had done a poor job of making its case against North Korea. There was no "Cuban Missile Crisis moment," in which Obama, like Kennedy fifty-two years before, presented his evidence. And what would he have shown? Everyone could recognize the Soviet missiles in Kennedy's spy-satellite photographs. But computer code was not made for vivid visuals.

"The best you can say is that there isn't a shred of evidence that anyone else was behind the hack," Kevin Mandia said to me at the time, after being called in to Sony to help.

In early January the White House announced some weak economic sanctions against North Korea—sanctions Kim may never even have noticed, given how many others had already been imposed. But afterward, the officials who designed that "proportional response" acknowledged that it was ridiculously weak, given the gravity of the attack on Sony and the president's vow that the United States would not tolerate intimidation. Part of the problem, some of them admitted, was that many Americans simply didn't believe the accusation against North Korea, or did not want to believe it. And the White House was still not ready to turn over evidence in public.

"If this had been a missile attack, we could have proved it," one frustrated senior official told me. "No one would doubt us. But cyber is a different thing."

In truth, even if he had been able to make the evidence public, Obama's options were limited. Economic sanctions, the first tool of presidents who want to show they are doing something without risking conflict, might have been satisfying on the day they were announced. But there was no evidence that sixty years of sanctions had slowed, much less stopped, the North Koreans' assembly of a good-sized nuclear stockpile. Why would sanctions do any better against a cyberattack?

In a series of Situation Room meetings in December, as aides tried to prepare options for Obama, more aggressive action was considered and rejected. The NSA and Cyber Command presented a list of responses Obama could order to disrupt North Korea's ability to connect to the outside world—one reason so many suspected America's fine hand when the North's Internet links through China went dead for a while. (It now looks likely that switch was pulled by the Chinese themselves.) But in a hint of things to come in dealing with Russia, Obama's aides feared that a counterstrike could start a cycle of escalation they could not stop. Clapper himself made the argument that the only deterrent that will work is a good defense, one that convinces would-be attackers that they will fail.

In short, in the Sony case, the government and the country got a glimpse of the disturbing, ambiguous nature of cyber conflict. It does not look like war as we know it, nor does it resemble Hollywood's depictions of a devastating cyberattack. The Sony attack demonstrated how profoundly a new generation of weapons has changed the geography of conflict between states. The new targets will likely be all civilian: even a movie studio, hardly critical infrastructure, makes for a ripe target.

"In the end," one of Obama's advisers told me with resignation in his voice, "we're a lot more vulnerable than they are."

CHAPTER VII

PUTIN'S PETRI DISH

In the twenty-first century we have seen a tendency
toward blurring the lines between the states of war and peace.
Wars are no longer declared and, having begun, proceed
according to an unfamiliar template.

—*Valery Gerasimov, chief of the general staff of the Russian Federation
Armed Forces, on Russia's hybrid warfare strategy, 2013*

I N THE LAST days of June 2017, Dmytro Shymkiv was 4,600 miles
from Ukraine, dropping off his kids for summer camp in upstate
New York. It was the family's annual summer break from life in Kiev,
a capital that still lives uncomfortably between the tug of old Soviet
culture and the lure of new Europe.

At camp, the kids could practice their English and learn what it's
like to be American teenagers. But for Shymkiv, a broad-faced entrepreneur with spiky hair, then forty-one, who became one of Ukraine's
most recognizable techies long before he was lured into government
service to help complete a revolution, the daily cyber battle with Moscow was never far away. Even in the mountains of New York.

"I had just gone out for a run," Shymkiv recalled later that summer
as we sat in his office in the presidential palace in Kiev, down the hall
from President Petro Poroshenko. "When I came back, I caught my
breath, and I looked at my phone and there was no real news. But then,

on social media, there were indications of a problem. And not a little problem."

Then the texts started pouring into his phone. Something—his staff could not tell exactly what—was freezing computers around Ukraine, simultaneously and seemingly permanently.

His first thought was that the Russians were back.

BEFORE SHYMKIV UNEXPECTEDLY found himself playing the role of four-star general in the world's most active cyberwar, he was a computer-obsessed kid growing up in a distant corner of the Soviet Union, and thinking constantly about how to get to the West. By the time he was a teenager, the Soviet Empire was no more, and by his twenties he had become one of the country's first tech entrepreneurs, before pivoting to lead Microsoft's small Ukraine operation. There he discovered just how vulnerable the country's backward technological foundation—full of old machinery and pirated, unpatched software—was to a massive cyberattack. He knew how simple it was for Russia to exploit Ukraine's weaknesses in the two wars simultaneously under way in Ukraine.

"There's a shooting war in the Donbass, since Crimea," he told me, referring to the eastern corner of the country where Russia's military forces were conducting a guerilla war against Ukraine, after Vladimir Putin ordered the seizure of the Crimean territory in early 2014. "And there is a digital war, every day, in Kiev." Shymkiv lived five hundred miles and a world away from that grim shooting war. But he had a front-row seat to the digital war, and it helped galvanize him to political action.

In February 2014, Shymkiv had taken vacation days from Microsoft to join the protests in Maidan Square at the center of Kiev—ground zero in the revolution that ousted Viktor Yanukovych, the corrupt former president and Russia-puppet. He camped with the protesters for two weeks, clearing snow and, ultimately, giving lectures on digital technology in the freezing cold in what came to be known, half in jest, as the "Open University of Maidan." He kept his Microsoft affiliation quiet; the company didn't know how the revolution was going to turn out and didn't want to be associated with the uprising. But Shymkiv broke his cover one night when Poroshenko—the opposition politician who would ultimately prevail and emerge as the country's president—came through. The two men chatted, a move that shocked some of Shymkiv's fellow protorevolutionaries. Yanukovych, of course, was spending millions of dollars to stay in power, relying on the advice and services of Paul Manafort, his friend and chief political strategist. Ultimately, though, he fled to exile in Russia. The election to replace him, in May 2014, amounted to a stark choice between a Ukraine that would surrender to Putin and one that Shymkiv and a generation of young Ukrainians imagined—a country that would turn to Europe. That election was a major target for Putin, who sought to defeat Poroshenko or, if that failed, cast doubt on his legitimacy and the integrity of the Ukrainian democratic process.

Thirteen more months would pass before Donald Trump glided down the golden escalator at Trump Tower to announce his candidacy for president of the United States. But for anyone looking for a preview of coming attractions, this was it.

Putin's cyber army went to work. Teams of hackers had scoped the Ukrainian election system, and planned their intrusions. On Election Day, they were ready. At the critical moment, they wiped out data in the system that tallied votes. But that was just the beginning. The hackers also managed to get into the reporting system that announced the results, altering the vote counts received by television networks. For a brief while, as news of the tally unfolded, it appeared to the Ukrainian media that Dmytro Yarosh, the leader of the nationalist and pro-Russia Right Sector Party, had emerged as the unlikely winner.

It was, of course, all a digital mind game. The Russian hackers didn't think the television declaration would stick. Rather, they simply sought to create chaos, and fuel an argument that Poroshenko manipulated the results to win. The plot failed: Ukrainian officials detected the attack, and corrected the results a nail-biting forty minutes before the networks aired them. Poroshenko had won, though not overwhelmingly—he had about 56 percent of the vote. Russia's own television networks, apparently unaware that the cyberattack had been detected, announced the phony results, with Yarosh as the victor.

Within weeks Poroshenko had contacted Shymkiv, whom he knew only vaguely beyond that encounter in the square. "He didn't give me much of a choice," Shymkiv later said with a laugh. Soon the guy who had started in computing by playing with the portable Sinclair computers of the '80s had been handed two tasks, both impossible: reforming Ukraine's corrupt institutions and securing the country against the daily cyber onslaught from Russia.

Now, three years later, in the woods of New York state near his kids' summer camp, Shymkiv fixated on his phone screen as texts from his Ukrainian colleagues pinged him in staccato. They reported that at around eleven-thirty in the morning computers across the country abruptly stopped working. ATMs were failing. Later the news got worse. There were reports that the automatic radiation monitors at the old Chernobyl nuclear plant couldn't operate because the computers that controlled them went offline. Some Ukrainian broadcasters briefly went off the air; when they came back, they still could not report the news because their computer systems were frozen by what appeared to be a ransomware notice.

Ukraine had suffered cyberattacks before. But not like this one. The unfolding offensive seemed targeted at virtually every business in the country, both large and small—from the television stations to the software houses to any mom-and-pop shops that used credit cards. Computer users throughout the country all saw the same broken-English message pop onto their screens. It announced that everything on the hard drives of their computers had been encrypted: *"Oops, your important files have been encrypted . . . Perhaps you are busy looking to recover your files, but don't waste your time."* It went on to make the dubious claim that if they paid $300 in Bitcoin, the hard-to-trace cryptocurrency, their data would be unlocked.

The attack was designed to look like a national shakedown scheme. It wasn't. The hackers weren't after money, and they didn't get much.

This was "NotPetya"—so nicknamed by Kaspersky Lab, which was itself suspected by the US government of providing back doors to the Russian government via its profitable security products. (The attack got its odd-sounding name because cyber-threat experts, trying to understand the inner dynamics of the attack, found elements in it that were similar to malware called "Petya" used in an attack the year before.) It didn't seem coincidental that the malicious code detonated just before the holiday that marks the adoption, in 1996, of Ukraine's first constitution after its break from the Soviet Union. But how had the hackers managed to freeze so many systems at once—upward of 30 percent of the nation's computers, of many different types?

It turned out that Ukraine's own backwardness—and an archaic remnant of its past—had played into the hands of the attackers. In true post-Soviet style, Ukraine required businesses to use a common piece of accounting software, M.E.Doc. It was clunky, it was old, but it was required by the state. Corrupting the software with malware was ridiculously easy: No one had invested in updating it in years. In fact, it used an outdated "platform" that had not even been supported by its manufacturer since 2013. No updates, no security patches.

By the time Shymkiv sped back to Kennedy Airport, his staff had discovered that the attack was no one-day event. "It turned out that bringing all those businesses down was the very end of a much bigger operation," he told me later. For months, the forensics showed, the Russian hackers had been gathering intelligence on Ukraine's top businesses, downloading emails and looking for everything from passwords to good blackmail material.

"Then, at the end, when they were done, they planted the bombs," Shymkiv said. "It was like the old Soviet days: First you rob the village, then you burn it."

· · ·

IT IS TEMPTING to think of cyberwar as something that takes place separate and apart from other conflicts, that what happens in the cloud is somehow divorced from what happens on the ground. When nations first built air forces, they thought something similar: dogfights in the air were one campaign, shooting in the trenches another. It was not until World War II that the concept of a "single battle space"—air, land, and sea—took hold. In some corners of the world that concept was already happening in cyber. It was just harder to see.

In the battle for Ukraine's territory and its soul, conventional war and cyberwar did more than just complement each other. They became the Möbius strip of twenty-first-century conflict, one continuous band, with surfaces that seem to blend seamlessly into each other. Putin showed the world how effective this strategy, what the Pentagon terms "hybrid warfare," can be.

The strategy was hardly a state secret. In fact, Valery Gerasimov, the chief of the general staff of the Russian Federation armed forces, described it in a much-quoted 2014 article in a Russian defense journal (the wonderfully named *The Military-Industrial Courier*) articulating what is now widely known as the Gerasimov doctrine.

Gerisamov described what any historian of Russian war fighting knows well: a battlefield war that merges conventional attacks, terror, economic coercion, propaganda, and, most recently, cyber. Each component enhanced the others. This blended approach had long helped Russia to project power around the globe, even when it was outgunned and outspent. Stalin was a master of information warfare, at home and abroad, and used it to increase his odds of victory in conventional war. If it confused and divided his enemies at home, all the better.

What is different now is the great amplifier of social media. Stalin would have loved Twitter. Skillful as he was as a propagandist, his transmission capability was primitive. "What's new is not the basic model; it's the speed with which such disinformation can spread and the low cost of spreading it," American political scientist Joseph Nye, the man who invented the term "soft power," wrote in describing how Russia was making use of "sharp power." If soft power is the ability to win over other societies because of the attractiveness of your culture, economy, and civic discourse, sharp power is the ability to insert the knife, stealthily and surgically. As Nye says, "Electrons are cheaper, faster, safer, and more deniable than spies."

There were many critiques of the Gerasimov doctrine, most arguing that far too much importance was given to a single article in a weekly journal. They contended that Gerasimov was simply observing an element of military strategy that both long pre-dated Putin and wasn't specific to Russia.

Fair enough, but Gerasimov's observations grew ever more relevant because cyber had forever altered the hybrid warfare game, and Russia had incorporated it more brilliantly than most other powers. When Gerasimov published his article in 2013, American war fighters looking at cyberpower were still focusing on its physical effects on power plants or equipment, as embodied by Operation Olympic Games. For them, cyberwar was one thing, information war another. To the Russians, it was all on a spectrum. At one end was pure propaganda. Then came fake news, manipulated election results, the publication of stolen emails. Physical attacks on infrastructure marked the far end.

Ukraine was where the techniques all came together, starting in early 2014. In the country's east, Putin sent in the Little Green Men— his unofficial army of soldiers, so named for their unmarked green uniforms—to maintain a simmering, low-level insurrection that used assassinations and bombings to keep the Ukrainian government off balance. To Putin, his plainclothes fighters served the same role on the streets that his hackers served on the Internet: deniability. Accustomed to an era when battles were fought by soldiers, the international community was hesitant to act without ascertaining the same level of attribution that insignia once provided. His green men and hackers alike were cloaked in enough ambiguity that Putin could get away with attacks consequence-free—even when there was little doubt that he was the source.

But Putin also recognized early, long before the West caught on to his scheme, that Ukraine's political divisions were ripe for exploitation. The divide between the Russian-speaking sections of the country and the rest would be particularly vulnerable to his cyber schemes, designed to hollow out a state, gradually degrade its institutions, and undermine confidence in everything from election boards to the courts and the embattled local governments. Not surprisingly, every technique Americans soon worried about began in the Ukraine: manipulated election results, fictional online personas who widen social divisions and stoke ethnic fears, and what was called "fake news" before the phrase was twisted into new meaning by an American president.

Putin's goals in Ukraine were as much psychological as physical. He wanted to declare to Ukrainians that their country exists only because Russia allows it to exist. Putin's message to the Ukrainians was simple: *We own you.*

It is no surprise that Putin picked the Soviet Union's old bread-basket for this experiment. Ukraine has never qualified for NATO membership. And even if it ever does manage to qualify, it is unclear whether NATO will take the risk of accepting it into the fold. Putin could attack without fear that the Western alliance will do more than issue international condemnations and sanctions. And in 1994, when Ukraine voluntarily gave up the nuclear weapons based there since the Soviet days—destroying them in return for a vague commitment that all nations will "refrain from the threat or use of force against the territorial integrity or political independence of Ukraine"—it also gave up any credible threat that it could strike back.

That commitment to Ukraine's independence—or what the West carefully termed an "assurance" because they actually committed to nothing—was shown to be empty when Putin seized Crimea in March 2014. The territory, he argued, had been part of Russia from 1783 to 1954, when Khrushchev handed it over to the Ukrainians. It was a blurry-enough history, and Putin rightly calculated that no American president or European leader would risk lives to defend a Russian-speaking corner of a faraway nation, especially outside the Western alliance.

In keeping with the Gerasimov doctrine, the violent seizure of the Ukrainian territory included political tactics, as Putin sought to boost the legitimacy of his actions through a "democratic" referendum over the status of the territory in March 2014. Media accounts suggested that the decision of the parliament to hold a referendum in the first place was achieved by fraud. One indicator of the dubious electoral practices, *Forbes* later reported, was that 123 percent of registered voters in Sevastopol cast ballots in the referendum.

There was enough confusion, plausible deniability, and disinterest in the region as a whole that Putin largely got away with it. At the time, he was doing the same in Syria, preparing the ground for what would become a full-scale military intervention in 2015. But the United States was oddly passive in both cases. Obama seemed fatalistic about Ukraine when he told Jeffrey Goldberg of *The Atlantic* that "the fact is that Ukraine, which is a non-NATO country, is going to be vulnerable to military domination by Russia no matter what we do." He was similarly cautious about Syria. When the Pentagon and the National Security Agency came to him with a battle plan that featured a sophisticated cyberattack on the Syrian military and President Bashar al-Assad's command structure, Obama said he saw no strategic value in pushing back in Syria.

In both cases, the United States and its allies deployed their standard tool for when military action seems too costly and doing nothing seems too feckless: economic sanctions. In the face of the Gerasimov doctrine and Putin's asymmetrical warfare, the best the US could do was make it difficult for Putin to ship out his oil and gas or lure new investors to revitalize a languishing Russian economy. After oil prices collapsed in late 2014, the sanctions began to cause real pain—chasing away foreign investors and undercutting Putin's support by undercutting growth. And one of the potential investors who was trying yet again to build a hotel in Moscow was Donald Trump.

During the first year of sanctions, one European diplomat who dealt often with Russia reported, "the Russians were telling the oligarchs, 'Just wait it out. The sanctions cost Europe too much business. They will go away.'" But in fact, the sanctions held, and in the United States they received overwhelming bipartisan support. Overwhelming—but not unanimous, at least after Donald Trump came along. One of the most striking aspects of one of the interviews Maggie Haberman and I conducted with Trump during his presidential campaign—long before there were charges that Trump was somehow in Putin's thrall—came when the new-to-foreign-affairs candidate told us he had doubts that the sanctions made sense at all. He did it in typical Trumpian style, assuring us of his deep interest in Ukraine and then asking why Americans should be paying the whole price for keeping Putin at bay:

> Now I'm all for Ukraine, I have friends that live in Ukraine, but it didn't seem to me, when the Ukrainian problem arose, you know, not so long ago, and we were, and Russia was getting very confrontational, it didn't seem to me like anyone else cared other than us. And we are the least affected by what happens with Ukraine because we're the farthest away. But even their neighbors didn't seem to be talking about it.
>
> And, you know, you look at Germany, you look at other countries, and they didn't seem to be very much involved. It was all about us and Russia. And I wondered, why is it that countries that are bordering the Ukraine and near the Ukraine—why is it that they're not more involved? Why is it that they are not more involved? Why is it always the United States that gets right in the middle of things, with something that—you know, it affects us, but not nearly as much as it affects other countries.

He then argued, "we're fighting for the Ukraine, but nobody else is fighting for the Ukraine."

"It doesn't seem fair," Trump told us, never lingering on what Putin was doing to the Ukrainian people or the offenses to the country's sovereignty. "It doesn't seem logical."

That was the part of the interview, we learned later, that the Russians noticed.

BEFORE THE UNITED STATES was worried about Russian meddling in American elections, its fears were a lot more basic: the potential for a "Cyber Pearl Harbor." That was the line Leon Panetta, by then Obama's defense secretary, had used in 2012, in a speech aboard a World War II aircraft carrier moored in New York Harbor. Such a strike, he told an invited audience, could "paralyze and shock the nation and create a new, profound sense of vulnerability."

He was hardly the first to use the evocative phrase; it had been employed for its rhetorical power for more than a quarter of a century. But Panetta, a savvy California politician, understood the power of the imagery. Congress, he once told me, "has a hard time funding defenses against a threat it can't see." So, even if it required bending reality a bit, he needed to liken cyberattacks to the most devastating surprise attack of the twentieth century. "We couldn't get Congress to focus on the issue," he told me later. "Somebody had to ring the bell, and the way to do that was to look at the potential for what a cyberattack could mean for our country."

Yet Panetta, better than almost anyone, understood why the analogy was imperfect. The most devastating cyberattacks, his experience taught him, were the most subtle. As CIA director, the post he held before he moved over to the Pentagon, he was a key player in Operation Olympic Games—and he had come to appreciate that much of the power in that attack came from its corrosive psychological effects as much as its destructive effects.

Panetta had reported to Obama in 2009 and early 2010 that the Iranians were dismantling parts of their enrichment center because of their inability to comprehend what was happening, and their fear of more calamities. In fact, even after Panetta delivered the news to Obama that the Stuxnet worm had gotten loose and was replicating itself across the globe, they agreed to keep the attacks under way for a while. The Iranians probably still hadn't grasped what was going on, Obama and Panetta bet, so the weapon had some lingering utility.

Panetta's real worry after he gave the speech in 2012 was less about an attack that had the drama of Pearl Harbor and more about one with the subtlety of Olympic Games. His staff spent endless hours mapping out what it would look like if an attack hit American industrial control systems—either quietly paralyzing America's ability to defend itself or causing damage akin to what the United States and Israel inflicted on the Natanz nuclear plant. When the electric grid began to fail, he thought, or communications were lost to submarines at sea, it might not look at first like a cyberattack. It might look more like a screw-up. Which of course is exactly how cyberattacks often unfolded in Ukraine, where screw-ups were a pretty common explanation for just about anything that went wrong.

ANDY OZMENT, of course, didn't need a wake-up call in the days before Christmas 2015. When he stepped into the Department of Homeland Security's giant war room—the National Cybersecurity & Communications Integration Center—it was clear that something was going wrong in Ukraine. The screens in the center mostly monitor events in the United States. But the command center also has linkages to the National Security Agency and Computer Emergency Readiness Teams, the organizations that keep each nation's networks running around the world. And everyone on those channels was buzzing about the electric outage in Ukraine, because in the cyber world, what happens in Kiev almost never stays in Kiev.

For more than a year, secret briefings inside the US government suggested the Russians were already implanting similar software in the United States. In chilling detail, they revealed the degree to which a foreign power was poised to turn out the lights. Ozment knew that the Russians, among others, were littering American power plants, industrial systems, and communications networks with implants that could be used later on to alter data or shut those systems down. Since 2014, intelligence agencies had been warning that Russia was likely already inside the American electric grid. The malware took many forms, often called "BlackEnergy."

The implants scared the hell out of American defense officials—but they were determined not to show it. In their most benign mode, the implants are useful for surveillance—broadcasting back to their home base news about what is happening inside a network. But what makes cyber threats different is that the same implant that is used for surveillance can be repurposed as a weapon. All it requires is the injection of new code. So on one day, the implant may be sending back blueprints of the electric grid. The next day it can be used to fry that grid. Or wipe out data. Or allow someone in a remote locale to take control of the equipment—and drive it off a cliff, so to speak.

The problem, as Ozment saw it, was that no one knew whether the Russians intended the hack to be a nuisance, an attack, a warning, or a rehearsal for something much larger. Perhaps they were just exploring how easy—or difficult—it was to get inside America's electric utilities, each of which is configured differently. Imagine a bank robber with global ambitions, facing the question of how to break into a series of bank vaults in New York, London, and Hong Kong. No two would be exactly alike. Everything would have to be custom-designed: Disabling the alarms, busting in, and making it impossible for outsiders to figure out what you are doing. A good escape plan, with no fingerprints or DNA left behind, would help too.

Naturally, when the Pentagon, the FBI, and utility executives looked at malware in an American power plant—especially the BlackEnergy exploit, which became widespread starting in 2014—their minds went immediately to the most extreme scenario: the Russians were preparing to shut down everything that makes America hum. But as Michael Hayden, the former NSA and CIA director, often said, "Breaking into a system is one thing, and breaking it is another." He was right: The Russians—and other nations—had been lurking inside American utilities, financial markets, and cell-phone networks for years. But so far they hadn't hit the kill switch.

In its own operations, the United States has also been cautious. Any destructive attack to actually *break* a foreign system requires many levels of approval, including from the president. There were looser rules about just entering a system and looking around—espionage instead of "preparing the environment" for attack. Yet as Martin Libicki, a cyber expert at the US Naval Academy, noted, to the country or company on the receiving end, that distinction may mean little: "From a psychological perspective, the difference between penetration and manipulation may not matter so much."

Perhaps that is why the Obama administration almost reflexively decided to treat the early breaches in the American utilities' networks as a classified secret. Senior members of Congress, selected staffers, and utility company CEOs were taken into locked, signal-proof rooms for briefings on the intelligence. There was no note taking. "It was ridiculous," one of them complained to me soon after. Under the strict rules they were given, the utility executives were barred from sharing the information with the people who administered their networks. Put another way, the only people who could do something about the problem—or at least prepare backup systems—were prohibited from knowing about it.

The intelligence agencies said they feared that if the discovery were made public it would tip off the Russians to the quality of our detection systems and perhaps to how deeply the NSA is into Russian systems as well. No doubt that was a risk. But could anyone imagine the United States withholding similar intelligence about an impending terror attack that might bring down a bridge, or blow up an electric substation? Almost certainly not—they would want everyone on alert. The rules were different for cyber.

For the Russians, there were no consequences in the Ukraine or in the United States for hacking into the grid. This situation would play out time and again: The NSA did not want to expose intrusions in American systems, for fear of exposing "sources and methods." The White House did not want to reveal what was known for fear that, as one of Obama's top advisers put it, "someone will then ask, 'What do you plan to do about it?'" Eventually, the Russian hack was exposed by private cybersecurity firms that detected the same malware in the utility grid that the government was seeing. And government officials would play coy, discussing the issue only in background conversations, rarely on the record. (Attacks by North Korea seemed to be the exception.)

Ozment knew all this recent history as he looked at what was happening in Ukraine. It made for several urgent questions: Was Ukraine a test run for something Russia was planning in the United States? Or was it simply part of the shadow war under way for two years in a faraway land?

No one really knew for sure.

THE CHRISTMASTIME ATTACK in Ukraine turned out the lights for only 225,000 customers, for a few hours. But Ozment suspected that switching off the grid, even briefly, was all the Russians intended. After all, this attack was about sending a message and sowing fear. It has never been clear, from what has so far been declassified, whether Putin himself knew about the Ukraine power attack in advance or if he ordered it. But whether Putin knew or not, the attack demonstrated in the cyber realm what the Russians had already demonstrated in the physical world by retaking Crimea: they could get away with a lot, as long as they used subtle, short-of-war tactics.

Ozment knew the United States had to understand how the Russians pulled off the attack. This was, after all, the first publicly acknowledged hack of electric utilities that actually turned out the lights. With the help of the Department of Energy and the White House, he assembled a team of experts—some of his own first responders and others from big electric utilities—and negotiated with Ukraine to dispatch them to Kiev. The instructions they carried were simple: go figure out what happened and whether the United States is vulnerable to the same kind of attack.

The team came back with a mixed answer. While the Ukrainians did not have defenses as sophisticated as many American utility companies, a quaint oddity in Ukrainian systems ultimately saved them from an even greater disaster. It turned out that their electric grid, built by the Soviets, was so antiquated that it wasn't entirely dependent on computers.

"They still had the big, old metal switches that ran the power grid back in the pre-computer age," Ozment explained, as if admiring the simplicity of an original Ford Model A engine. The investigators reported that Ukrainian engineers got into their trucks and went scrambling from one substation to another, looking for switches that they could throw to route around the computers and turn the lights back on.

Score one for a creaking, antiquated system, particularly since it would take months for the Ukrainians to rebuild their damaged computer-based network controls. But Ukraine's resilience was not much comfort to the Americans who read the reports and thought about their own vulnerabilities. Few American systems still had these rusting old switches—they were eliminated long ago. And even if the American utilities had hung on to the old systems, the engineers who knew how they worked had long since retired. It would, one energy executive told me, be quite a challenge to find someone with the knowledge of how to save the day.

The reports contained some other lessons. Ozment's investigators had discovered a trail of evidence in Ukraine that pointed to careful planning, a professional job. Before the attackers struck, he said, they had been inside Ukraine's electrical grid for more than six months. They had stolen passwords, slowly gained administrator privileges, and learned everything they needed to take over the system and lock out the control-room operators.

They had also left bits of computer malware waiting to explode, like land mines inside the network. Just as in the Sony attack, a "KillDisk" program had been used to wipe out hard drives—turning the computers at the utilities into a useless pile of metal, plastic, and mice. For good measure, a "call center" for customers to dial to report an outage was flooded with automated calls, to maximize frustration and anger.

In the dry wording of one after-action report, "the outages were caused by the use of the control systems and their software through direct interaction by the adversary." Translation: Not only had the computer systems been attacked, they had been controlled from afar, likely outside Ukraine's borders. It was a metaphor for what Russia wanted to do with the whole country.

Read closely, the reports underscored how much the Russian attackers seemed to learn from what the United States and Israel did in preparation for the Stuxnet attack on Iran's nuclear program. Each of the steps seemed familiar—the patience, the care that went into mapping the systems and making the losses so hard to recover from. One night I pressed a former official who had delved deeply into the attack. "They followed the American script," he agreed. "And they figured out how to use it against us." It wouldn't be the first time, he said, nor the last.

Nothing about the Ukraine attack was particularly sophisticated technically, Ozment observed, "but it is dangerous to confuse sophistication with effectiveness." That was the message being circulated back in Washington, where, as private firms noted, some of the malware found in the Ukrainian power grid was the same "BlackEnergy" code that was also in the US power grid. BlackEnergy hadn't turned the lights out in Ukraine. But it helped prepare the operation.

Ozment paused a moment. "When you looked at what happened," he concluded, "it was pretty chilling."

THE RUSSIANS WEREN'T done in Ukraine. Other attacks were still building steam. Russia pummeled Ukraine in 2016—a year when

President Poroshenko declared that the Ukrainian government had been hit by 6,500 cyberattacks in just two months, though these were mostly harassment rather than serious attacks. The message was clear: the cyberattacks were part of a continuing, low-level conflict meant to keep the Poroshenko government off balance. Clearly Ukraine was being tested. The Russians wanted to determine if there were any limits. They discovered none.

What happened in Ukraine confirmed the corollary to the Gerasimov doctrine: As long as cyber-induced paralysis was hard to see, and left little blood, it was difficult for any country to muster a robust response. The attack would make a lot of news, but it would be unlikely to galvanize much action, especially if it were unclear for a few weeks who was responsible. Meanwhile, the government under attack could appear both helpless and hopeless.

Putin's gamble was clearly paying off. He had sent a powerful signal that there were ways to take the conflict to the Ukrainian capital and undermine the government of President Poroshenko without sending a single tank into the city. And, on the world stage, these cyberattacks showcased Putin's newest tools—which other countries had not figured out how to deter, or respond to if deterrence fails.

Incapable of striking back at Russia, Ukraine remains locked in the oddest kind of just-below-the-radar conflict, one intended to keep the country from straying too far toward Europe. For Putin, the country still serves its centuries-old role as a buffer zone with the West. The daily digital air raids are intended to keep the country in perpetual instability—an onslaught that everyone soon got accustomed to, a feature as permanent as Saint Sophia Cathedral in the middle of Kiev.

So the Ukrainians shrugged when the Russians hit the power grid again, in December 2016. That attack was briefer, but it hit the capital. And it showed that the Russians were learning. In 2015 they had gone after a distribution system; when they came back they had gone after one of Kiev's main transmission systems. And when a company called Dragos unpacked the code, they found a new kind of malware, called "Crash Override," that was designed specifically to take over the equipment in the grid. It was based in part on artificial-intelligence techniques, enabling it, as Andy Greenberg of *Wired* wrote, to "launch at a preset time, opening circuits on cue without even having an internet connection back to the hackers." It was the cyber equivalent of a self-guiding missile.

It would work almost anywhere, with a few tweaks.

Shymkiv played down its import when he saw me in his office, about seven months later. But he admitted that Ukraine was "the petri dish for every cyber technique the Russians want to trot out." What he lacked was a game plan for stopping those attacks; he was always playing defense.

It turned out he wasn't the only one without a strategy for dealing with a newly aggressive Russia. Washington didn't have one either.

CHAPTER VIII

THE FUMBLE

I cannot forecast to you the action of Russia. It is
a riddle wrapped in a mystery inside an enigma;
but perhaps there is a key.
—*Winston Churchill, October 1939*

I N THE MIDDLE of 2015, long before the 2016 presidential primaries
heated up, the Democratic National Committee asked Richard
Clarke, a hard-bitten national security fixture in Washington, to assess
the political organization's digital vulnerabilities.

Clarke was best known as the counterterrorism chief in the Clinton and Bush national security councils, who had given warning that
Osama bin Laden was planning a massive attack on the United States.
He was the one who, in the aftermath of the September 11 attacks,
famously told the relatives of victims that "your government failed
you," and blamed President Bush's White House for ignoring the many
warnings he had issued. Embittered by his government experience but
very much a DC creature, he went off to start a cybersecurity firm,
Good Harbour International.

He wasn't surprised when the Democratic National Committee
called. "They were an obvious target," Clarke told me later. But he
was amazed when his team discovered how wide-open the committee's

systems were. As it stood, the DNC—despite its Watergate history, despite the well-publicized Chinese and Russian intrusions into the Obama campaign computers in 2008 and 2012—was securing its data with the kind of minimal techniques that you might expect to find at a chain of dry cleaners.

The committee employed a basic service to filter out ordinary spam, but it wasn't even as sophisticated as what Google's Gmail provides; it certainly wasn't a match for a sophisticated attack. And the DNC barely trained its employees to spot a "spear phishing" attack, the kind that fooled the Ukrainian power operators into clicking on a link, only to steal whatever passwords are entered. It lacked any capability for anticipating attacks or detecting suspicious activity in the network— such as the dumping of data to a distant server. It was 2015, and the committee was still thinking like it was 1972.

So Good Harbour came up with a list of urgent steps the DNC needed to take to protect itself.

Too expensive, the DNC told Clarke after the company presented the list. "They said all their money had to go into the presidential race," he recalled. They told him they'd worry about the security issues after Election Day. That response came as no surprise to anyone who knew the DNC as a bailing-wire-and-duct-tape organization held together largely by the labors of recent college graduates working on shoestring budgets.

Of the many disastrous misjudgments the Democrats made in the 2016 elections, that one may rank as the worst.

"These DNC guys were like Bambi walking in the woods, surrounded by hunters," a senior FBI official told me. "They had zero chance of surviving an attack. Zero."

WHEN AN INTELLIGENCE report from the National Security Agency about a suspicious Russian intrusion into the computer networks at the DNC was tossed onto Special Agent Adrian Hawkins's desk at the end of the summer of 2015, he was already overwhelmed. Whenever the FBI was called in to investigate a major cyber intrusion in Washington that had struck a think tank, law firm, lobbyist, or political organization, the assignment usually ended up on Hawkins's desk. As an agent making his way up the ladder at the Washington Field Office, Hawkins possessed the weary air of a man who had seen every variety of hack. And he had, from espionage to identity theft to attempts at destroying data.

So when the DNC hack got tossed onto his growing pile, it did not strike him or his superiors at the FBI as a four-alarm fire.

"It was hard to find a prominent organization in Washington that the Russians *weren't* hitting," another veteran of the FBI's cyber division told me later. "At the beginning, this just looked like espionage. Everyday, ordinary spying." They assumed that the DNC intrusion was probably just another case of hyperactive Russian spies looking to buff their résumés by bringing home a bit of political gossip. After all, the DNC was not exactly where you went to find the nuclear codes.

In September, Hawkins called the DNC switchboard, hoping to alert its computer-security team to the FBI's evidence of Russian hacking. He quickly discovered they didn't *have* a computer-security team. He ended up being transferred to the "help desk," which he found singularly unhelpful. Then, someone on the other end of the line handed the phone to a young information technology contractor with no real experience in computer security. His name was Yared Tamene.

Hawkins identified himself on the line and explained to Tamene that he had evidence that the DNC had been hacked by a group that the federal government (but no one else) called "the Dukes," a Russia-affiliated group. He didn't go into details about their long history of breaking into other government agencies or how stealthy they were in avoiding detection. He couldn't—much of that was treated as classified information, even though private security firms had published extensively about the group, which most called "Cozy Bear."

Tamene jotted down some information about how to identify the malware. Later he composed an internal DNC memo and emailed it to his colleagues. "The FBI thinks the DNC has at least one compromised computer on its network and the FBI wanted to know if the DNC is aware, and if so, what the DNC is doing about it," Tamene wrote. And he went back to his workday tasks.

Of course, the DNC wasn't aware. And it wasn't doing anything about it.

Perhaps Tamene's sangfroid in the face of Hawkins's news was due to the fact that he had no memory of Watergate. Or perhaps it was because he wasn't a full-time employee at the DNC; he worked for a Chicago-based contracting firm they'd hired to keep their computers operating. He was in charge of keeping the network running, not keeping it safe. Most important, he thought that Agent Hawkins might be spoofing him, that perhaps the call was from someone impersonating an FBI agent. So when Hawkins left a series of voice messages a month later, Tamene didn't call back. "I did not return his calls," Tamene later wrote to his colleagues, "as I had nothing to report."

It wasn't until November 2015 that the two spoke on the phone again. This time Hawkins explained that the situation was now worsening. One of DNC's own computers—it wasn't clear which one—was transmitting information out of its headquarters. In a memo Tamene later wrote, he said Hawkins specifically warned him the machine was "calling home, where home meant Russia."

"SA Hawkins added that the FBI thinks that this calling home behavior could be the result of a state-sponsored attack," Tamene wrote. Implicit in the memo was this reality: The FBI might see the DNC's data flowing outside its building, but it didn't have the responsibility to protect privately owned computer networks. That was the job of the DNC itself.

That second warning should have set off alarms—but there is no evidence that it did. Tamene's information never made it to the DNC's top leadership—the committee was then run by Debbie Wasserman Schultz—or so they insisted later. And the FBI, for its part, was focused elsewhere, including on the mysteries of Hillary Clinton's Chappaqua computer server. With no one treating the issue with urgency, the ridiculous dance of cues and miscues between Hawkins and Tamene continued, extinguishing the last, best chance to halt the biggest political hack in history.

ANYONE LOOKING FOR a motive for Vladimir Putin to poke into the election machinery of the United States doesn't have to look far: revenge.

In December 2011, Russia had just completed a parliamentary election that all observers, foreigners and Russians alike, believed was riven with ballot-tampering and fraud. For the first time since Putin had come to power, protesters poured into the streets. The chants said it all: "Putin is a thief," and "Russia without Putin!"

Putin, of course, had won the election, but not by much. His party, United Russia, lost a lot of ground; it was barely hanging on to its majority after three smaller parties ate away at its numbers. United Russia surged at the end of the vote count, raising everyone's suspicions. Golos, the only independent election monitoring group in the country, discovered that its website had been attacked so that it could not report suspicious activity, and that was after a court had fined it for violating the law by publishing accounts of campaign abuses. Monitors from the Organization for Security and Cooperation in Europe, who had been invited to observe, reported blatant ballot stuffing, no surprise to Russians who had seen a YouTube video of an election commissioner filling out ballots as fast as he could. As the video went viral, Russians took again to the streets.

Enter Hillary Clinton, in her third year as secretary of state, who by then had realized her ill-executed effort to stage-manage a "reset" with Russia was doomed from the moment she handed her Russian counterpart a giant button that had mistranslated the word "reset."

Several days after the voting, she issued a bland State Department declaration about the election, right out of the standard State playbook. "The Russian people, like people everywhere, deserve the right to have their voices heard and their votes counted," she said, never mentioning Putin or his party by name. Repeating the boilerplate language of generations of secretaries of state, she later spoke out about the United States' "strong commitment to democracy and human rights," and particularly "the rights and aspirations of the Russian people to be able to make progress and to realize a better future for themselves." There was no threat of sanctions.

Clinton and her aides did not think they were saying anything particularly unusual; calling out Russia for antidemocratic behavior was standard stuff. Putin, however, took the declaration personally. The sight of actual protesters, shouting his name, seemed to shake a man known for his unchanging countenance. Then, in typical style, he smelled an opportunity. He declared that the protests were foreign-inspired. And at a large meeting he was hosting, he accused Clinton of being behind "foreign money" aimed at undercutting the Russian state.

"I looked at the first reaction of our US partners," he went on, with barely contained anger. "The first thing that the secretary of state did was say that [the elections] were not honest and not fair, but she had not even yet received the material from the observers."

"She set the tone for some actors in our country and gave them a signal," Putin said. "They heard the signal, and with the support of the US State Department began active work." The implication wasn't exactly subtle: the United States and its Russian stooges—not Putin himself—had rigged the election. Putin's charges may have been designed to draw attention from his own election meddling, but he quite cleverly tapped a vein of Russian conspiracy-mongering when it came to American foreign involvement. The United States did not exactly have clean hands when it came to influencing elections in other countries. Italy and Iran were notable targets for CIA election manipulation and coup-organizing in the 1950s, and Putin would cite American efforts to kill Castro in Cuba and to mount covert influence campaigns for elections in South Vietnam, Chile, Nicaragua, and Panama. He argued that the pro-Western colour revolutions in Georgia, Kyrgyzstan, and Ukraine in the early 2000s, as well as the Arab Spring, similarly arose from soil tilled by the United States and fertilized with American cash. "Put your finger anywhere on a map of the world," Putin said in 2017, "and everywhere you will hear complaints that American officials are interfering in internal election processes."

Putin's moral equivalence didn't hold much water. While in the bad old days the CIA would have brought bags of cash to Italian politicians and Chilean strongmen, election influence had since become the territory of the State Department, whose techniques were significantly more timid and transparent. When the United States intervened in contemporary elections, it usually did so to assure that more people had access to the vote. Rather than cash, it stuffed suitcases with an "Internet in a box" to defeat crackdowns on information. It sent out "consultants" to teach novice candidates how to campaign, helped build independent courts, and of course monitored election fraud.

But Putin would respond that the United States tried to oust Hamid Karzai in Afghanistan, and he would be right. He argued that the Americans merely wrapped their election-influence operations with flowery descriptions about "democracy promotion."

Not surprisingly, Putin quickly put down the 2011 protests and made sure that there was no repetition in the aftermath of later elections. But the mix of personal grievance at Clinton and general grievance at what he viewed as American hypocrisy never went away. It festered.

PATIENT ZERO IN the new Russian campaign to strike back at the State Department and Hillary Clinton was Victoria Nuland.

The granddaughter of Orthodox Jews who emigrated to the Bronx after escaping Stalin's rule, "Toria," as she was often called, never wandered far from her Russian heritage, or from the bruises it left on the family. Her view of Putin's Russia was no secret: The only thing the former KGB officer understood, she argued, was tough pushback.

"I didn't mind attempting a reset; all administrations try that," she told me. "But it had to be a reset with no blinders on." Putin, in her view, was a superior tactician. He had a spy's sense of opportunism, particularly if he could play the spoiler. But he was far weaker when it came to long-term strategy. So when he poked—with military action, a cyberattack, intimidation—he needed a sharp poke back. Let him get away with something, and he would only take more.

Nuland cemented that view as she began to move into the upper echelons of the State Department. Many members of her Foreign Service class quickly learned how to file the burrs off their speech as they described American interests. Nuland, though, made no effort to hide her realpolitik view of what the United States needed to do to defend its interests, and when she wasn't on the State Department podium she might lace her views with a few well-chosen epithets. If one was wandering the halls of the State Department looking for someone to make the strongest case for diplomacy backed by the threat of force, her office was always a good place to start.

Putin was highly aware of her role. Early in her career, she had worked for Strobe Talbott, Bill Clinton's deputy secretary of state and an old friend of both Clintons. Talbott was a Russia hand, and Nuland's hawkishness was often useful in giving his outreach to Russia some bipartisan appeal. That became clear in later years: after all, not many foreign service officers who rose in the Clinton camp made it into Dick Cheney's inner circle—Nuland was his deputy national security adviser—and went on to become a favorite of Barack Obama. For all his caution, Obama was impressed by Nuland's willingness to get in Putin's face.

Before Obama was elected, Nuland and Putin already had developed an unhappy, if distant, history. As American ambassador to NATO during Bush's second term, she pressed the allies to resist Russia's nascent efforts to move from cooperation to confrontation. That was an uphill push. Many in NATO believed the narrative that Russia, its economy roughly the size of Italy's and its population shrinking—could not afford to take on Europe and the United States. "Toria is breaking a lot of china at NATO," an ambassador from another NATO ally told me during a visit to the alliance's headquarters at the end of her tenure. "Most of it needs to be broken."

By the time of the 2011 Russian parliamentary election, Nuland had returned to Washington and was serving as the State Department spokeswoman under Hillary Clinton. Putin may have thought Nuland was behind Clinton's decision to call out the voting fraud. He had good reason for the suspicion: it was Nuland's job to denounce the election fraud from the State Department's podium.

But by early 2014, when the Maidan revolution was in full swing in Ukraine, Nuland had moved on from the podium and was the assistant secretary of state for European and Eurasian Affairs, and the point person for the Ukraine crisis for a new secretary of state, John Kerry. After years in the field, she knew all the players, but she also knew the stakes for Putin and for Russia should Ukraine prove capable of resisting the Russian tractor beam.

"If countries like Ukraine really can elect their own leaders," she said, "if young people can really say and do whatever they want, if the country gets rich by knitting into Europe rather than being a big gas station, then the Russian people themselves will look and say, 'We'd like more of that.'"

In her usual type-A style, Nuland was working the phones constantly during the protests, trying to negotiate a way for a peaceful settlement that would ultimately get President Yanukovych out of power. While Yanukovych turned to Paul Manafort in a desperate bid to remain in office, Nuland was trying to broker a new election. Getting there would require putting together a coalition government between Yanukovych's party and the opposition, but the opposition politicians so distrusted the Ukrainian strongman that they would not negotiate without a neutral observer. Nuland thought that role should be played by the European Union. The Russians, of course, wanted the entire process upended—it could only lead to trouble for their chosen puppet, Yanukovych.

One weekend, in the midst of the Ukraine crisis, Nuland was home in Virginia discussing the dilemma of getting to an election over the phone with Geoffrey Pyatt, the American ambassador to Ukraine. They ran through the question of who might serve in a Ukrainian government if Yanukovych were pushed out of office, and how they might privately urge some of the opposition leaders to serve.

The conversation turned to how the European Union was balking at its role as a neutral observer, declining to name someone to fill the role.

Nuland is not known for her patience for diplomatic dithering.

"Fuck the EU," she told Pyatt.

"Exactly . . ." he responded.

Nuland's mistake, of course, was that she and Pyatt were speaking over an open, unencrypted line. No surprise there—the State Department's secure phones were perpetually broken. Nuland fully understood the risk, and she was pretty certain the Russians were listening in. But she was undeterred. She and Pyatt were talking in private about a strategy they had already described in public. Maybe it would be a good thing, she thought, if Putin's henchmen reported back that she meant business.

The conversation ended, one of many in a series of urgent phone calls as the crisis played out on Kiev's streets. Then, two weeks later, the audio, edited, suddenly appeared on YouTube, highlighting her "Fuck the EU" line. Nuland and Pyatt, alongside many others, were stunned. "They had not dumped a phone call on the street in twenty-five years," she told me later. This was a new tactic. Only later would Nuland discover that she had been the canary in the coal mine. The YouTube video represented a new Russia, determined to exploit new techniques.

Nuland later said she had been expressing a "tactical frustration" with her European allies. Yet the tape was edited so that it didn't sound that way. Instead, her words seemed to signal a breach in the relationship between the United States and the EU, exactly the kind of split Putin would love to exploit.

The uproar that followed was predictable. Pyatt, a Californian who had come to the Foreign Service in 1989 via Yale, wondered if his career were over. (It wasn't—he went on to be ambassador to Greece.) In Washington, Nuland spent a lot of time apologizing. Though she was one of the State Department's top diplomats, she also briefly wondered about her job, until a few days later when she attended a state dinner at the White House and, seeing President Obama, repeated her apology directly to him.

When he smiled and said, in a low voice, "Fuck 'em," a clear reference to the Russians, she knew she was OK.

THE BROADCAST OF the Nuland-Pyatt phone call marked a turning point for Russian "active measures." The public release of the recording was just the start. As the year wore on Russia kept pouring non-uniformed troops into parts of Ukraine, and accompanying the surge with what Gen. Philip Breedlove, the NATO commander, called "the most amazing information warfare blitzkrieg we have ever seen in the history of information warfare." Ukraine and other states, he urged, needed plans to launch counter-propaganda efforts, and perhaps counter-cyberattacks.

Breedlove knew that NATO was totally unprepared. It had hesitated to enter the cyber age. While decades ago NATO had worked out elaborate plans to use nuclear weapons in defense of Europe, even storing some near its headquarters in Brussels, it had no cyber counterattack unit, nor any expertise in "information warfare." While visitors were often taken to a giant, gleaming computer security center, it was designed to protect only its own networks. And until a few years ago, one senior American told me, it protected those networks only during the weekday.

"No one had budgeted to get 24/7 monitoring, even of our most sensitive networks," he said, shaking his head. "The only thing they forgot to do was send a postcard to the Kremlin with a note that they could save themselves a lot of hard work if they just attacked NATO on nights and weekends."

THE ONE TIME I encountered Yevgeny Prigozhin, the man who would work to alter the 2016 election, he was not surrounded by trolls, and his employees were not creating bots. It was May 2002, on a river in Saint Petersburg, and he was serving dinner to George W. Bush and Vladimir Putin.

In retrospect, Bush's trip was the high point in the relationship between Washington and Moscow. It was the Texan politician's first trip to Russia as president. In Moscow, the first stop, the two men had just signed a nuclear arms–reduction treaty. Bush called it a moment to "cast aside old doubts and suspicions and welcome a new era," and that seemed like it might still be possible.

I caught only a glimpse of "Putin's Chef," as he became known, in the few minutes a group of correspondents traveling with Bush were escorted in and out of Prigozhin's floating restaurant, one of the city's most fashionable dining spots, so that they could describe the scene to readers. I thought he was just a chef. Was I ever wrong.

A decade and a half later, Prigozhin re-emerged an oligarch. It wasn't a bad landing spot for a guy who'd spent his youth in jail and started his culinary career running a hot-dog stand. Before the 2016 election had heated up, he already stood accused of cooking up a far larger project for Putin: a propaganda center called the Internet Research Agency, housed in a squat four-story building in Saint Petersburg, Prigozhin's hometown. From that building, tens of thousands of tweets, Facebook posts, and advertisements were generated in hopes of triggering chaos in the United States, and, at the end of the process, helping Donald Trump—a man who liked oligarchs—enter the Oval Office.

Stalin would have been proud of the Internet Research Agency. It existed in plain sight and yet was not what it appeared. It was not an intelligence agency, yet it learned some of their skills. It was nothing fancy: a building, filled with young talent willing to spend twelve hours a day peddling pure fiction, some of it aimed at the Russian market, some aimed at Europe. The best talent was assigned to the American desk, stuffed with some of the agency's highest-paid, most imaginative writers. Fake news didn't come cheap.

Stalin used Soviet propaganda to recruit Americans, undermine capitalism, and sow fear and distrust. The Internet Research Agency did the same, but Facebook and other social media sites gave it reach Stalin could scarcely have imagined.

It is still unclear if the idea behind the IRA came from Prigozhin, Putin, or someone in between. But its creation marked a moment of profound transition in how the Internet could be put to use. For a decade it was regarded as a great force for democracy: as people of different cultures communicated, the best ideas would rise to the top and autocrats would be undercut. The IRA was based on the opposite thought: social media could just as easily incite disagreements, fray social bonds, and drive people apart. While the first great blush of attention garnered by the IRA would come because of its work surrounding the 2016 election, its real impact went deeper—in pulling at the threads that bound together a society that lived more and more of its daily life in the digital space. Its ultimate effect was mostly psychological.

There was an added benefit: The Internet Research Agency could actually degrade social media's organizational power through weaponizing it. The ease with which its "news writers" impersonated real Americans—or real Europeans, or anyone else—meant that over time, people would lose trust in the entire platform. For Putin, who looked at social media's role in fomenting rebellion in the Middle East and organizing opposition to Russia in Ukraine, the notion of calling into question just who was on the other end of a Tweet or Facebook post—of making revolutionaries think twice before reaching for their smartphones to organize—would be a delightful by-product. It gave him two ways to undermine his adversaries for the price of one.

It may be years, if ever, before there is any clear understanding of how large a role Putin himself played in developing and executing "active measures" for the Internet age. He is not known as a user of social media himself. But he had a KGB alumnus's appreciation of its power.

As start-ups go, the Internet Research Agency (called Glavset by the Russians) rose pretty fast. By sometime in 2013, it was getting its foothold in Saint Petersburg and began hiring. Soon it operated on a multimillion-dollar budget whose source is still murky. It quickly employed not only news writers but graphics editors and experts in "search engine-optimization" to ensure the greatest reach for its pro-Russian messages. And it took advantage of the fact that Facebook did little, at least at the time, to determine whether a member was really a person—or just a bot. The whole strategy depended on convincing other users that a fake persona was real. The IRA's hackers were essentially playing the same role as those soldiers who had shed their uniforms in Ukraine.

And the digital little green men took the propaganda battle to the enemy's territory. The American campaign began in September 2014 with text messages like the one aimed at the residents of St. Mary Parish, Louisiana, warning of a toxic fume release from a chemical plant. It turned out that "Columbia Chemical," the plant, reported to be in the throes of the accident, did not exist. But the fear was palpable. Then came the echo-chamber rumors that Ebola was running wild in parts of the United States, stoked by Russian trolls who set about amplifying the rumors on social media—spreading the hashtag #EbolaInAtlanta with fake news and video accounts of the incident.

In its first headquarters, a blocky, four-story building in Saint Petersburg, at 55 Savushkina Street, the Agency's dozens of twenty-somethings learned to "troll" critics of Putin and journalists who delved too deeply into what the agency did. It didn't take them long to perfect the art form. As Putin and his Chef had learned, it is easy to make a critic miserable in the Twitter age. And the Internet Research Agency performed the task well—growing to 80 employees with outsized influence online.

In late 2014, the agency dug into its social media campaign to commence its disruption of the US elections. The group deployed hundreds of fake accounts on Facebook and thousands on Twitter to target populations already divided by issues like immigration, gun control, and minority rights. These were early, "beta" efforts—propaganda on the cheap. All it required was figuring out how best to game the algorithms that fed Facebook news feeds, or fueled retweets on Twitter.

Then they moved on to advertising. Between June 2015 and August 2017, investigators later discovered, the agency and groups linked to it spent thousands of dollars on Facebook ads each month—at a fraction of the cost for an evening of television advertising on a local American television station. The reach was surprisingly broad. In that period, Putin's trolls reached up to 126 million Facebook users, while on Twitter they made 288 million impressions—seemingly stark numbers given that there are about 200 million registered voters in the US and only 139 million voted in 2016. But it is unknowable if they had much impact.

Putin's trolls posed as Americans or fake American groups on social media and promoted clear messages. Their Facebook posts might feature a doctored picture of Clinton shaking hands with Osama bin Laden or a comic depicting Satan arm-wrestling Jesus. "If I Win Clinton Wins," the Satan figure says. "Not if I can help it," the Jesus figure responds. (Users were encouraged to "Like" the image to help Jesus triumph, which in turn generated the Internet buzz needed to increase the picture's visibility based on Facebook's algorithm.) The purpose of the hundreds of posts like these, suggests Ryan Lizza's reporting in the *New Yorker,* was to "overwhelm social media with a flood of fake content, seeding doubt and paranoia, and destroying the possibility of using the Internet as a democratic space."

Yet social-media savvy could take the Russians only so far. In order to meddle in the United States they needed a better understanding of American electoral politics. The agency dispatched two of their experts—a data analyst and a high-ranking member of the troll farm, Aleksandra Krylova and Anna Bogacheva—to the United States, where they spent three weeks touring purple states: California, Colorado, Illinois, Louisiana, Michigan, Nevada, New Mexico, New York, and Texas—while another operative scoped out Atlanta. Along the way, they did rudimentary research and developed an understanding of swing states, a concept for which there was no parallel in Russian politics. The information that these agency researchers gathered during their weeks in the United States helped the Russians develop an election-meddling strategy based on the importance of purple states to the electoral map. That allowed the IRA to target specific populations within these states that might be vulnerable to influence by social media campaigns operated by trolls across the Atlantic.

In mid 2015—having mastered the art of social-media meddling—the trolls tested out a new tactic organizing a live event in the United States, according to an investigation by the Russian business magazine *RBC*. Using Facebook accounts based in Saint Petersburg, they posed as Americans and lured users to a free hot-dog event in New York. Of course, the trolls in Saint Petersburg didn't provide the promised food to the New Yorkers whom they watched gather through a publicly accessible webcam in Times Square; rather, it was a successful experiment proving that, from their screens in Russia, they could orchestrate events in the physical world. This seemingly small feat would soon move far beyond hot dogs, and into the realm of inciting clashes among rival American groups at political rallies in the "purple states" that the Russians were learning about. The magazine reported: "From this day, almost a year and a half before the election of the US President, the 'trolls' began full-fledged work in American society."

The use of Facebook events would evolve quickly. The following year the trolls recruited an actress to attend a rally for Trump in West Palm Beach dressed as Hillary Clinton in a prison uniform. She was paraded in a cage that was built by other Americans. They apparently didn't know that they were being paid by Russians in Saint Petersburg.

. . .

THE INTERNET RESEARCH AGENCY was not the only "proxy" force that was stepping up its game against the United States. So were the hackers working for Russia's several, often competing, intelligence agencies. And before they ever broke into the DNC, the most sophisticated team, the one working for the SVR—a descendant of the old KGB—had been focused on two juicy, high-value targets: the State Department and the White House.

The first strike was against the State Department's unclassified email system. (Like most government agencies, State maintained both a "high side" classified network and a separate "low side" network to communicate with the outside world.) It was a classic operation where the Russians inserted malware that created a link to their own command-and-control server abroad. When State Department staffers clicked on the "phishing" emails that the Russians had created, some purporting to be from American universities, the hackers were in. They could then copy emails at their leisure, hoping they might pick up some gossip, maybe a little policy debate, maybe an affair they could use for blackmail material.

With luck, they may have also found clues about how to get into the "high side" systems—the classified systems. By the time Kevin Mandia and his firm's experts came to look at the system, "the Russians were all over it," he recalled. They had gone after specific, high-ranking officials, including, of course, Toria Nuland. What Mandia saw in the State Department's system was an attack that was far more brazen than anything the Russians had attempted before. "They were just a lot more stealthy," he said.

Rumors of some kind of intrusion into the State Department systems had swirled in Foggy Bottom for weeks. The first hint I caught of the severity of the Russian hack came during a trip to Vienna on the third week of November 2014. A group of us, traveling with Secretary of State John Kerry, had landed in the city for another round of negotiations with Iran over its nuclear program. As I emailed and called

American officials, a seemingly innocuous email message popped up from the State Department. Forget trying to get ahold of members of the offices of public affairs, much less Kerry's negotiating team, by email over the weekend, it warned reporters. The entire State Department system was coming down for "system maintenance."

Eyes rolled. To anyone who had heard the rumors of a Russian intrusion, this reeked of a cover story. The real issue was obviously not maintenance, though the creaky system that connected the nation's diplomats together seemed at times little better than two paper cups and a string. This sounded more like standard operating procedure for damage control: to conduct a digital exorcism and flush out intruders, the first thing you had to do was bring the system down.

That wasn't going to be easy. By this point getting the Russians out of the State Department systems had already proved too difficult for the Department of Homeland Security. They had called for reinforcements from the NSA, on the theory that it takes a cyber thief to catch one.

Rick Ledgett, the man who had handled the Snowden investigation, suddenly found himself overseeing the operation to oust the Russians from the State Department networks. And he was cautioning that it had to be done right. He knew from bitter experience that while it was tempting to rush the ouster of a cyber invader, that was usually how you made mistakes. (The Navy learned this the hard way, when Admiral Rogers was the head of Fleet Cyber and Iranian hackers got inside their networks. The hackers were thrown out before all the implants they had put in the system were discovered, and soon they came back.) So the NSA experts started by identifying where the Russians were in the system, and where they had placed implants and a command-and-control center. Only then could the system be brought down, the invaders disconnected, and a new system raised, phoenixlike and hopefully with better security, in its place.

"These guys were really dug in," Kevin Mandia later told me. "And they weren't planning on leaving. Usually, you shine a light on the malware, and the guys at the other end scatter like roaches.

"Not the Russians. They had a point to make."

With some effort, a State Department team, backed up by Mandiant, the FBI, and the NSA, eventually chased the Russians out of the system. But it turned out they had just moved on.

No sooner had the battle at the State Department begun to wrap up than the Russians turned up a mile away—inside the White House servers. "The State Department was just winding up," Ledgett said, "and the White House was ramping up." Once again, the attackers hit the unclassified "low side," not the "high side" systems that run on different computers.

The exorcism process began all over again. As they had at the State Department, the Russians made clear that, having started their White House tour, they had no intention of leaving the premises. In the White House system, it turned out that the NSA and its partners had walked into something of a digital ambush. The Russians were mounting the attack from command-and-control servers they had placed around the world, to help hide their identities. Every time the NSA's teams of hackers cut the links, they found that the White House computers began communicating with new servers. No one had seen anything quite like it—a state-sponsored group of hackers in a digital dogfight.

To the NSA's team, it looked a bit like a video game with real-life consequences. "They seemed to be having a good time living in the White House system," one of the American officials noted, somewhat ruefully.

Later, Ledgett described what managing that battle was like, without ever mentioning that the hackers were from Moscow. "We saw for the very first time," he said, that "instead of disappearing, [the hackers] fought back. And so it was basically hand-to-hand combat in a network where we would take an action, they would then counter that."

The NSA, he said, would "remove their command-and-control channel to the malware, to the code that they were running," and the Russians "would counter that by introducing a new command-and-control channel."

In retrospect, it was also a new moment on the tactical battlefield of cyberspace, he said, "a new level of interaction between a cyberattacker and a defender."

Ledgett alluded to the fact that the NSA had a secret weapon of its own: It was "able to see them teeing up new things to do. So if you're the defender and you see what the adversary's gonna do," he added with some understatement, "then that's a really useful capability."

He appeared to be referring to some quiet help from the Netherlands. The tiny nation's intelligence agencies, according to an investigation by two Dutch news organizations, had penetrated a university building off Red Square in Moscow, from which the group of Russian hackers sometimes called "Cozy Bear" operated. But the Dutch hadn't just gotten into the computer systems, they also got into the building's security cameras. "Not only can the intelligence service now see what the Russians are doing," the Dutch report said, "they can also see who is doing it."

The Dutch alerted their intelligence liaisons in The Hague, and soon a link was created so that American intelligence agencies could see who was going in and out in real time. Those pictures were then fed into facial-recognition software so that it was possible to identify who was operating the computers.

Suddenly everyone—from the NSA to the FBI, and the White House Communications Agency—were caught in the usual dilemma when they identify invaders in their networks. Do they watch them, track their activities, maybe feed them some false information? Do they move quickly to throw them out? And were the Russians really seeking information, or did they want to get caught so they could learn about American detection capabilities?

And most important, at least to the Russians: Was Obama willing to escalate a confrontation, or dismiss it as another moment in the endless spy-versus-spy games that both sides played?

· · ·

In the end, the Americans won the cyber battle in the State and White House systems, though clearly, as events played out, they did not fully understand how it was part of an escalation of a very long war.

The battle for control of the computer networks at the State Department and the White House raised two big questions. First, why did the Russians choose to take on the United States so directly? And second, why did the Obama administration try to keep the whole series of incidents secret, including the hacks on the Pentagon and Congress?

The answer to the first question seems simple: The Russian hackers were strutting their stuff for the same reason Russian generals parade their tanks and missiles just across the border from Lithuania. It was the 2014 equivalent of what fighter pilots used to do in the Cold War, flying to the edge of Soviet-controlled airspace to see what would happen as the Russians scrambled their fleet.

"They were making it clear they were here to stay, and had the stuff to go up against our best," a senior intelligence official said to me. "There was lots they were doing that was still hidden—like the election hack—but they wanted us to know they had entered the big leagues."

But the bigger mystery was Obama himself. Once again he had chosen not to call the Russians out. At one Situation Room meeting, he told his intelligence officials that this was "just espionage." And if the United States was sloppy enough to let it happen, then the answer was to up our game on defense rather than think about retaliation.

When this computer game was over, the Russians had retreated— though tactically, and not for long. They had moved up the learning curve fast. It's unclear whether one could say the same of the Obama administration.

Remarkably, even with the hacks of the State Department and the White House in recent hindsight, in the waning days of 2015, no one told senior White House officials about the major, Russian-led intrusion into the DNC. Nor did the leadership of the DNC know: Special Agent Hawkins later told FBI officials that he was hesitant to email anyone inside the committee for fear of tipping off the Russians. In the first few months of phone-tag with Tamene, he never bothered to take a twenty-minute lunchtime walk from the Washington Field Office to the DNC headquarters, which had long ago moved from the Watergate to far blander quarters on Capitol Hill. "We are not talking about an office that is in the middle of the woods of Montana," said Shawn Henry, the former head of the FBI's cyber division, whose company was ultimately called in to help investigate. "We are talking about an office that is half a mile from the FBI office that is getting the notification."

It was a stunning lack of judgment all the way around—a failure to grasp the gravity of an old threat wrapped in a new technology. It was also the beginning of a series of fumbles across the board that undercut America's ability to react at a crucial point in time when it could have made a difference.

WHILE THE AMERICANS dithered, the Russians feasted. The communications failures between the DNC and the FBI gave Putin's hackers what they needed most: time. Exposed but not yet thwarted, they had the luxury of exploring every nook and cranny of the DNC's main server, which was only slightly larger than a laptop computer. Once this was drained, they moved on to targets outside the DNC.

Finally, by March 2016, six months after the initial calls, Tamene and his colleagues had met the FBI at least twice and now seemed convinced that Hawkins was, in fact, an FBI agent.

By then, it was too late. The Russians had already moved on to stealing the emails of the officials of the Clinton campaign itself.

Clinton had set up shop in Brooklyn and had a lot more money than the DNC. Remembering that Chinese hackers had broken into both Obama's and John McCain's 2008 campaigns, the Clinton team brought in some serious cybersecurity expertise. The result was that the campaign's own networks repelled several attacks, none of them wildly sophisticated. But the Russian hackers had a bigger game in mind: personal email accounts, where people tend to put complaints about the boss, their worries, and their contemplation of future personnel moves, along with documents that they don't want on a corporate network.

Top of the Russians' list was Clinton's campaign chairman, the taut, wiry John Podesta. There was no better-connected Washington insider. He had served as Bill Clinton's chief of staff. He had organized many campaigns. And he had deep, substantive knowledge on everything from climate change to cyber privacy—a topic on which he produced a report before leaving the Obama White House in 2015.

His familiarity with all things digital didn't help him much on March 19, 2016. That was the day a fake message, ostensibly from Google, showed up in his personal inbox. It warned him that someone was trying to break into his personal account. As it turned out, this phishing email came not from the Dukes but from a new team of Russian-affiliated hackers. The group succeeded with a similar phishing message to campaign aide Billy Rinehart, but Podesta was a far richer target.

Because Podesta was so focused on fund-raising and message-sharpening for the Clinton campaign, a handful of his aides managed his email for him. When the spear-phishing email arrived, declaring that he had to change his password, it was sent to a computer technician for a judgment on its legitimacy.

"This is a legitimate email," Charles Delavan, a Clinton campaign aide, replied to his colleague—the aide who had first noticed the phony alert. "John needs to change his password immediately." Delavan later told my *Times* colleagues that his bad advice was a result of a typo: He knew this was a phishing attack because the campaign was getting dozens of them. He meant to type that it was an "illegitimate" email, an error that he said has haunted him since.

And so, the password was changed immediately. Suddenly the Russians obtained access to sixty thousand emails, stretching back a decade.

CHAPTER IX

WARNING FROM THE COTSWOLDS

*Russia talk is FAKE NEWS put out by the Dems,
and played up by the media, in order to mask the big
election defeat and the illegal leaks!*
—@realDonaldTrump, February 26, 2017

*I never said Russia did not meddle in the election, I said "it may be
Russia, or China or another country or group, or it may be a
400 pound genius sitting in bed and playing with his computer."
The Russian "hoax" was that the Trump campaign
colluded with Russia—it never did!*
—@realDonaldTrump, February 18, 2018

I N THE SPRING of 2016, Robert Hannigan was eighteen months into his job as director of GCHQ—Britain's equivalent to the NSA—and he was getting accustomed to the rituals of the job. His past service to the government had been radically different: seeking peace in Northern Ireland under Prime Minister Tony Blair and adjudicating among bitterly competing British intelligence agencies at 10 Downing Street. But then he had been sent to one of those agencies, the Government Communications Headquarters, the blandly named bureaucracy that was still living off its reputation as the agency of brilliant oddballs who had cracked the German codes with the Enigma machine during World War II, and saved Britain.

Hannigan's job was to bring GCHQ into the twenty-first century, the century of cyber conflict. Past heads of GCHQ barely communicated with the public, but on his first day on the job Hannigan took a direct shot at Silicon Valley firms in a column in the *Financial Times*. "However much they may dislike it," he wrote, "they have become the command-and-control networks of choice for terrorists and criminals," and must learn how to cooperate with the intelligence agencies of the Western democracies. Yet once he settled into the job, he found a player who worried him more than Facebook and Google: Vladimir Putin.

Hannigan thought Putin was causing a "disproportionate amount of mayhem in cyberspace." His staff of thousands of code breakers, signal-intelligence officers, and cyber defenders had soon learned to place the raw evidence of that mayhem atop the pile of intelligence they brought him each day, culled from their own piles of intercepted computer messages and phone calls.

On this particular day, around Easter in 2016, a series of messages plucked out of the Russian networks stood out.

In the inartful terminology of the digital world, it was mostly "metadata," Hannigan's staff told him. To Hannigan's frustration, he could not see its actual content. But it was clear that the traffic was controlled by one of Russia's premier intelligence agencies, the GRU, the aggressive military intelligence unit whose activities GCHQ tried to monitor around the clock.

What struck Hannigan, though, was where the messages appeared to have originated: the computer servers of the Democratic National Committee.

WHEN HANNIGAN SORTED through the message traffic, pausing to examine what would turn out to be a historic intelligence intercept, he was deep inside "The Doughnut," the Brits' affectionate name for the bizarre, round Cheltenham headquarters of GCHQ. From the air, the building actually looked more like a spaceship, as if aliens had decided to drop in on the quaint pubs of the Cotswolds: Stow-on-the-Wold and

Bourton-on-the-Water, the Shakespearean-era villages just down the road. The Doughnut's design was very Silicon Valley; once inside the secure zone, everyone worked in the open, cross-pollinating ideas.

Of the thousands of communications GCHQ intercepted every week or so, more and more from Russia were pulled out and placed atop the daily pile on Hannigan's desk. Like the CIA and NSA, British intelligence agencies had been surprised by the speed and stealth of Putin's annexation of Crimea in 2014. NATO nations were worried enough about stepped-up Russian bomber and submarine runs along the European coast—something they had not seen since Soviet days—that they had to devote more resources to tracking them all.

"We had gotten pretty complacent about Russia," one of Hannigan's national security colleagues told me. "There was still this overhang from the '90s that somehow the Russians would come to their senses and join the West and become our economic partners. Even when they attacked Georgia in 2008, people shrugged it off. It took a long time for reality to set in."

The Baltic states on Russia's edge now appeared, in the British official's words, a "vulnerable gray zone" that Putin would seek to destabilize. Soon after arriving at GCHQ at the end of 2014, Hannigan began pressing for more intercepts, more "implants" in the networks to which Britain had unique access, one of the last benefits of a dismantled British Empire. Every day came a torrent of new material: messages fleshing out Russia's support for the Syrian government of Bashar al-Assad, its maneuvers off Finland, its submarine runs.

To Hannigan, it was all new and fascinating. His background wasn't in intelligence; it was in the intersection of politics and national security. At first glance, he was easily mistaken for the very model of the polished British bureaucrat: buttoned down, with the perfect pedigree for a job that was all about discretion. To one of his aides inside the Doughnut, Hannigan's best attribute was a "puckish sense of humor about the ridiculousness of much of what we do in the intelligence business."

Though Hannigan was no intelligence professional, he was put atop

GCHQ because David Cameron, the prime minister, had come to rely on his judgment after years at 10 Downing Street. Already, Hannigan had broken a lot of china at the hidebound and overly secretive agency. The agency was born after World War I as the "Government Code and Cypher School," which pretty well defined its role in the twentieth century. Hannigan was born twenty years after World War II had ended, and it was his job to push GCHQ to figure out its role in the cyber age. It had survived since the glory days of Enigma at Bletchley Park, decoding messages and intercepting calls, but in a new era when defense and offense had blended, merely intercepting conversations was not enough.

So Hannigan began reorganizing GCHQ's structure and moving it beyond its roots in signals intelligence. He realized that, like the NSA, GCHQ needed to up its game in cyber skills—specifically "network exploitation" and "network attack." Month by month, Hannigan tried to push the agency into the future. On his watch, GCHQ scraped ISIS recruiting messages off their servers around the world. Hannigan particularly enjoyed seeing transcripts of ISIS cyber lieutenants fuming that they could not get into their own recruiting and communications channels.

Cheltenham, on the edge of the Cotswolds, is a place of splendid isolation, and with his family remaining in London, Hannigan had plenty of time to dig deep on the Russia intercepts. The one containing DNC data was a particular mystery.

"It didn't tell us much," he recalled. "It told us there was an intrusion, and something had been taken out of the committee. But I had no way of knowing what."

As Hannigan looked at the intercepted Russian communications from the DNC, it was his sense of history that made them stand out. He was only seven years old when the Watergate scandal broke, barely aware of the headlines from across the Atlantic. But he had become enough of a student of history and politics at university to immediately grasp the import of what the Russians seemed to be doing. "The DNC meant something to me," he said. "And it was an odd target."

It was unclear what they were looking for. The DNC wasn't a place to get military secrets, or even much policy. It was essentially a place to redistribute cash to campaigns. The goal was a mystery.

Hannigan thought his American counterparts needed to see these intercepts, and fast. He looked at them once more and asked his staff to be sure to flag them for the National Security Agency. This shouldn't get lost in the daily pile, he told them. This was sensitive stuff, and his American counterpart, Admiral Rogers, and his colleagues at the NSA, needed to know about it.

A few weeks later, Hannigan recalled, he received an acknowledgment "from someone senior" on Rogers's NSA staff. They appreciated the heads-up.

It was the last he heard from them about it.

INSIDE THE NSA, officials hint that they already had a pretty good idea of what the Russians were up to at the DNC, and they say the British were not the only foreign intelligence service to see evidence of the hack. But they were the most important, and that should not be surprising. For reasons of history, geography, and faded empire, GCHQ's access to the networks that feed into and out of western Russia are among the best of the "Five Eyes"—the five English-speaking victors of World War II who share the burden of intelligence gathering and most of what they harvest.*

Hannigan describes the Five Eyes as more of a club than a tightly run organization. It was, he said, a "signals intelligence creation dating from World War II, when Roosevelt and Churchill took a political decision to share their most sensitive cryptological secrets."

"I think Americans would be surprised by how many British experts we keep at the NSA," one senior British official with deep experience

---

* In addition to the United States and Britain, the other members are Canada, Australia, and New Zealand.

said to me a few years before the Russia investigation broke. "And I know the British would be surprised how many Americans are deep in our system."

In fact, the tie between the NSA and GCHQ was so tight that each placed its own officers in the other's headquarters, so they were partners rather than anonymous analysts on each end of the line. Snowden documents revealed that in Bude, on the southwest coast of Britain, there were 300 GCHQ analysts and 250 Americans in 2012, working on two projects—"Mastering the Internet" and "Global Telecoms Exploitation"—that picked up terabytes of Facebook entries, emails, phone calls, Google Maps searches, and histories of who visited what websites, and when. It was all legal, the British maintained after the operation was revealed, but the analysis section was based in Britain for a reason: there was more legal leeway than in the United States.

For obvious reasons, no one will be very precise about how the British picked up the traffic that led back to the DNC. But there are several clues. The Snowden documents reveal that GCHQ was plugged into two hundred fiber-optic cables, and could process information from forty-six of them simultaneously. That is quite a feat, since cable traffic runs at ten gigabits per second. The content of that traffic is mostly encrypted. But the British were able to pick up the metadata.

British access to the cables came courtesy of two leaders who were quite definitely of a pre-cyber age: Queen Victoria and President James Buchanan. When HMS *Agamemnon* and the USS *Niagara* met in the mid-Atlantic in 1858 to splice together the first copper cable, the queen and the beleaguered president used the new undersea line to transmit telegrams to each other. Britain then became the critical hub—the "termination point"—for even more cables laid across Europe and into Russia. "Termination points" are where the cables come ashore. And in both the United States and Britain, the intelligence agencies paid "intercept partners"—like AT&T and British Telecom—to keep teams of technicians at the termination site to mine and hand over data. The whole arrangement is ruled by court orders on both sides, kept secret to avoid blowback for the firms. Post-Snowden, the rules governing the system got a lot stricter. But the intelligence was also getting more valuable.

One hundred and sixty years later, the copper cables have been replaced by fiber-optic cables, which are more durable, higher-capacity, and harder to tap, and more than 95 percent of network traffic moves through them. One termination point in Cyprus, leaked documents showed a few years ago, has long been a particular bonanza for intelligence agencies. So has another in Asia, not far from North Korea. When Gen. Keith Alexander, then the head of the NSA, visited the Menwith Hill Station in Yorkshire in 2008, he asked, "Why can't we collect all the signals all the time? Sounds like a good summer project for Menwith."

He could have said something similar at other listening posts around the world, which are divided up for monitoring among the Five Eyes. While the Brits focus on Europe, the Middle East, and western Russia, the Australians monitor East Asia and South Asia—which is why operations in Afghanistan are often run out of Pine Gap, in the Australian desert. New Zealand owns the digital traffic in the South Pacific and Southeast Asia. Canada peers deep into Russia and covers Latin America. The United States, with huge collection budgets, looks at hot spots, starting with China, Russia, Africa, and parts of the Middle East. Naturally, such monitoring is a subject officials in each of those countries won't discuss openly, even years after the Snowden revelations.

One reason is that these termination points are no longer just a place to plug in headphones. They have become a way to inject implants—malware—into foreign networks. "Once they were all about defense," a telecommunications expert told me. "Today, they are also about offense."

They are also a huge risk, as the steady flow of global communication depends on them. If six or so termination points were blown up or seized, information flow in the United States would slow to a trickle. Phone conversations would halt, markets would be disabled, news would stop. "It's a tremendous vulnerability," one British official told me. "And a great opportunity."

So it was no surprise that Facebook and Google started laying their own cables.

IT WAS A sign of the Russian hackers' professionalism that they did not rush the stolen Podesta emails into public view after they obtained them in March 2016. Instead, they took their time sorting through the material, looking carefully at what might be especially valuable, such as Clinton's speeches to Goldman Sachs. She had refused to reveal the texts publicly, but here they were, in the stolen trove. (It turned out the speeches sounded a lot like the ones she used to give for free when she was Secretary of State.) The Russian strategy was one of patience: there would be a moment to reveal the contents of the emails, when they could do maximum damage.

At the DNC, Yared Tamene still saw no reason to be alarmed. He wrote in a memo on April 18 that a "robust set of monitoring tools" had finally been installed at the DNC—in other words, they had decided to pay for a burglar alarm.

Only later in April did Tamene, using those new tools, find evidence that someone had stolen credentials giving them access to all of the DNC's files. He called the DNC's chief executive, Amy Dacey, with news that there had been a major, recent breach and the DNC had probably lost most of its files—far more than they ever lost in the Watergate break-in.

Belatedly, panic set in.

FAR FROM WASHINGTON, another element of the Russian enterprise was playing out in Texas, Florida, and New York—all in plain sight.

While Russian intelligence agencies were hiring hackers to break into the DNC, the trolls and bot creators at the Internet Research Agency in Saint Petersburg were kicking into overtime. Paychecks had risen to $1,400 a week, a small fortune by Russian standards, especially for twentysomethings. In return, they worked twelve-hour shifts, churning out Facebook posts that hit on themes conveyed to them by email. On one floor, Russian-language trolls fought off opposition to Vladimir Putin. On another floor, they looked for any divisive issue in American society where a wedge could be driven via the Internet, to widen the natural fault lines in American politics and society.

Texas seemed particularly ripe for meddling. Few of the trolls and bot makers had been there, but they had read about it online and seen it in the movies. It didn't take much of a leap of imagination to form a "Heart of Texas" group that appeared to be based in Houston, but was actually operating near Red Square. They promoted a rally called "Stop Islamization of Texas," as if there were much Islamization to worry about. Then, in a masterful stroke, the Russians created an opposing group, "United Muslims of America," which scheduled a counter-rally, under the banner of "Save Islamic Knowledge." The idea was to motivate actual Americans—who had joined each of the Facebook groups—to face off against each other and prompt a lot of name-calling and, perhaps, some violence.

It was a testament to how easy it is to mislead some subgroups of American citizens on the web with a few cheap bots and someone imitating a local resident. But no one was more amazed than the young Russians in Saint Petersburg, who, their own emails later showed, could not believe their targets were so gullible.

IF YOU ARE going to catch a Russian inside your networks, hiring a Russian who thinks the way the attackers think isn't a bad idea. By that measure, Dmitri Alperovitch was the right man for the job.

In his mid-30s, with sandy hair and broad smile, he was already a fixture in the Washington firmament: a cyber specialist who was a regular at foreign policy forums and seemed as interested in the geopolitics of the business as the bits and bytes. But it was hardly preordained that Alperovitch would get so far.

He was the son of Soviet nuclear scientist Michael Alperovitch, and spent his childhood and early teen years in Moscow, in the waning days of the Soviet Union. In 1986, when Dmitri was about five years old, Michael narrowly escaped an assignment that would have left his son fatherless. A fire had broken out at the Chernobyl nuclear power plant, and panicked Soviet officials wanted Michael and his colleagues to check it out. Michael had a bad feeling and declined. The scientists who went all developed cancer and died soon after.

His life spared by good luck, Michael began to think it might be time to get out. His opportunity came shortly after the breakup of the Soviet Union. The Alperovitch family left Moscow in 1994, moving first to Toronto before settling in Chattanooga when Michael had landed a job at the Tennessee Valley Authority. Dmitri eventually enrolled at Georgia Tech, graduating with what was, at the time, a rare degree in cybersecurity.

Out of college, Alperovitch bounced around a number of the digital stations of the cross, eventually joining McAfee, known for its early virus-protection products. His job was to analyze state sponsors of cyberattacks, and he did it well, publishing a long paper about a China-based group called "Shady Rat," which was behind the theft of intellectual property from American companies. McAfee had been acquired by Intel, the country's leading chipmaker, and the paper took off as one of the best-researched pieces of work tying the Chinese government to what Keith Alexander, then the head of the National Security Agency, used to call the "greatest transfer of wealth in history."

Unsurprisingly, the Chinese didn't care much for the research. Suddenly they were showing up in Intel's offices in Beijing, inspecting business licenses—completely unrelated to Alperovitch's work, naturally. One day, he recalled, he got a call from one of the company's top executives. "Do you realize we do 60 percent of our business in China?" he remembers the executive asking.

Actually, he hadn't known that. He resigned the next week and in 2011 moved on to create the cybersecurity firm CrowdStrike with entrepreneur George Kurtz. Alperovitch knew how to follow the bits. His partner knew how to manage the law-enforcement landscape.

It was good timing; the Russians were coming.

"WHY DON'T YOU come up and we'll do a little health check?"

That was the seemingly benign invitation that Shawn Henry—a former FBI cyber expert whom CrowdStrike had recruited to serve as their chief security officer and president of their information security team—received from Michael Susman that April. Susman had prosecuted cybercrimes for the Justice Department, then moved to Perkins Coie, a law firm that counted both the Hillary Clinton campaign and the DNC among its clients.

CrowdStrike was accustomed to such calls, and soon their forensic engineers were tapped into the computers at the DNC, scanning them for signatures of known bad actors in cyberspace. Reams of data began flowing back to Henry and Alperovitch.

It took less than a day to find what they were looking for, but the full result was startling. It was at that moment that they discovered the DNC had been hacked by not one Russian intelligence group but two. And both had left plenty of fingerprints.

Alperovitch and his colleagues had long before nicknamed the first group "Cozy Bear," the one the FBI referred to as "the Dukes." It was a play on the Bear nicknames of the Cold War era. (Others called the group "APT 29" for "advanced persistent threat.") Cozy Bear was the first group to infiltrate the DNC, the evidence suggested, the one Hawkins had seen when he first called the committee.

It wasn't until March 2016 that "Fancy Bear," a competing Russian group associated with the GRU, the military intelligence unit, broke into the computers of the Democratic Congressional Campaign Committee before moving into the DNC networks as well. That was the hack that Robert Hannigan's spies at GCHQ had detected. Fancy Bear probably didn't know that the SVR-linked Cozy Bear group was already there. At least, that was Alperovitch's theory.

"These guys are deeply competitive with each other," he told me. "They want approval from Putin, they want to say 'Look what I did!'" And Fancy Bear was clearly busy—they were the ones sorting through Podesta's email trove.

Once it was clear where the invaders were coming from, Alperovitch threw himself into the investigation. The mystery was what the Russian groups planned to do with the information they had stolen. As Alperovitch noted dryly to me one day, "No one expected what this turned out to be."

ALPEROVITCH KNEW WHAT he needed to do at the DNC: replace its entire computer infrastructure. Otherwise, he would never know for sure where the Russians had buried implants in the system.

For the six weeks after CrowdStrike moved into the DNC headquarters, it worked quietly to prepare for a total replacement of the committee's hardware, making the usual excuse that there were maintenance operations under way. Then, on one weekend in late spring, everything was shut down. DNC employees were told to turn in their laptops and phones for a "system upgrade."

"There were people who thought this was a front for layoffs," since the DNC was perpetually broke, Alperovitch recalled. They were relieved to discover that their jobs were safe, but when they got the equipment back the next week, the hard drives had been wiped clean and new software installed.

By now the DNC leadership had moved from total ignorance to total panic. They began meeting with senior FBI officials in mid-June, fully nine months after Agent Hawkins had been switched to the help line. Babies had been conceived and born in the time it took the DNC, and the US government, to wake up. Now the debate was over whether to make public what was going on.

The motivation of the DNC and its chairwoman, Debbie Wasserman Schultz, seemed clear: She wanted to gain a bit of sympathy for the Democrats, who had been attacked by the Russians, and put

Donald Trump on the spot, since he had been nothing but complimentary about Putin. In mid-June, the DNC leadership decided to give the story of the hack to the *Washington Post*. It would leak soon enough anyway, they thought.

The *Post* ran with it, but it was a sign of how little thought was being given to Russian manipulation at the time that, as we played catch-up in the *Times* newsroom that day, it was difficult to get much interest in the story from the editors managing coverage of the strangest presidential campaign of modern times. At that moment, a few Russians mucking in the DNC didn't exactly seem like a repeat of Watergate. The story was buried deep in the political pages.

The Obama administration also had a difficult time getting excited. They resisted demands from the DNC that the government do a quick "attribution," as they had in the Sony case, and have the intelligence community publicly name the Russians as the offenders. The FBI said its own investigation was being hindered by the DNC, which it still viewed as being less than fully cooperative; the DNC would not allow the FBI access to its main servers, so the FBI was getting evidence secondhand, from CrowdStrike.

The government's reluctance to "attribute" the hack to the Russians was hardly unusual. There was always concern in the intelligence agencies about revealing sources and methods. And while it was one thing for a private security firm like CrowdStrike to name the Russians, the US government had to have a much higher level of certainty. "If you do it," one senior intelligence official said to me, "you have to be prepared to answer the question, 'So what are you going to do about it?'"

Susman, the lawyer for the DNC, thought that the government's argument was pretty ridiculous; CrowdStrike didn't need secret sources to figure this out, and the Russians had not exactly hidden their tracks. "You have a presidential election under way here and you know that the Russians have hacked into the DNC," he recalled saying at one meeting with DNC executives and their lawyer. "We need to tell the American public that. And soon."

The day after the *Post* and the *Times* ran their stories, though, it became clear that the Russians had a larger plan.

A persona with the screen name Guccifer 2.0 suddenly burst onto the web, claiming that he—not some Russian group—had hacked the DNC. His awkward English, which became a hallmark of the Russian effort, made it clear he was not a native speaker. He contended he was just a very talented hacker, writing:

> Worldwide known cyber security company CrowdStrike announced that the Democratic National Committee (DNC) servers had been hacked by "sophisticated" hacker groups.
>
> I'm very pleased the company appreciated my skills so highly))) But in fact, it was easy, very easy.
>
> Guccifer may have been the first one who penetrated Hillary Clinton's and other Democrats' mail servers. But he certainly wasn't the last. No wonder any other hacker could easily get access to the DNC's servers.
>
> Shame on CrowdStrike: Do you think I've been in the DNC's networks for almost a year and saved only 2 documents? Do you really believe it?

Guccifer 2.0 offered a few DNC documents, which he advertised as just a sampling of a vast trove. They included a lengthy piece of opposition research prepared by the DNC as they struggled to understand Trump, with chapter headings like: "Trump Is Loyal Only to Himself" and "Trump Has Repeatedly Proven to Be Clueless on Key Foreign Policy Issues." There was also a chart listing major donors to the DNC, where they lived, and how much they had given.

"And it's just a tiny part of all docs I downloaded from the Democrats' networks," he wrote, adding that the remainder, "thousands of files and mails," were now in the hands of WikiLeaks.

"They will publish them soon," he predicted.

It was clear that morning that the hack was not simply about campaign intelligence gathering. It was intended to be the cyberattack equivalent of broadcasting the conversation about Ukraine between Victoria Nuland and Geoffrey Pyatt. There was only one explanation for the purpose of releasing the DNC documents: to accelerate the discord between the Clinton camp and the Bernie Sanders camp, and to embarrass the Democratic leadership. That was when the phrase "weaponizing" information began to take off. It was hardly a new idea. The web just allowed it to spread faster than past generations had ever known.

Anyone who had followed the Russian hacking groups knew that there was little chance that Guccifer 2.0 was simply a savvy, lone hacker. But the name he chose was a clever play: It was taken from "Guccifer," the screen name of a Romanian hacker who was then sitting in jail, after famously breaking into the email accounts of former Secretary of State Colin Powell and former President George W. Bush.

It didn't take long for online sleuths to puncture the tale and point to evidence that Guccifer 2.0 was far more likely a committee of hackers somehow linked to the GRU, the Russian military intelligence unit. Lorenzo Franceschi-Bicchierai, who wrote for *Vice*, had the inspired idea of sending Guccifer 2.0 a direct message. He got an instant answer: Guccifer 2.0 said he was Romanian.

So Franceschi-Bicchierai used Google Translate to ask Guccifer 2.0 some questions in stilted Romanian. The answers came back in equally stilted Romanian. It quickly became clear that Guccifer 2.0 didn't speak the language; he was using Google Translate too. A deep look at the documents he was posting showed they had been written in a Russian version of Microsoft Word, and were edited by someone who identified himself as Felix Edmundovich. That name seemed a tip of the hat to the founder of the Soviet secret police, Felix Edmundovich Dzerzhinsky. (Dzerzhinsky Square in Moscow, where the KGB headquarters was located, got renamed after the fall of the Soviet Union, but Dzerzhinsky would soon have a bit of a revival.)

The more Franceschi-Bicchierai conversed online with Guccifer 2.0, the more he became convinced that he was dealing with "a group of people" who were not very skilled at covering their tracks. In fact, they didn't really seem to want to cover them. And another outlet for the documents suddenly appeared: "DC Leaks," a site established just a few months before, but not active until the end of June. It was another indication that making selected stolen documents public was part of a larger plan, one that had been formulated months in advance.

By THE TIME Donald Trump arrived in Cleveland, Ohio, in the third week of July 2016 to accept the nomination of a Republican Party still stunned by his rise, questions about his campaign's connections to Russia were already in the air. The millions of dollars that Paul Manafort, Trump's campaign chairman, made in Ukraine on behalf of the now-exiled, pro-Putin former president of the country was under growing scrutiny—which would lead to his resignation, and eventually his indictment. The digital break-in at the DNC was strange enough, but Trump's insistence that there was no way it could be definitively traced to the Russians was even stranger.

As I arrived in Cleveland, though, the biggest mystery seemed to be Trump's own refusal to say anything remotely critical of Russia, and especially of Vladimir Putin. Every other Republican candidate for president I had covered—Bob Dole, George W. Bush, John McCain, Mitt Romney—had gone out of their way to stress their suspicions of Russia's motives, and particularly Putin's.

Yet Trump kept declaring he admired Putin's "strength," as if strength was the sole qualifying characteristic of a good national leader. In an interview with Fox News he refused to say if he had ever spoken to Putin. That seemed odd, because he was also attempting to make the case that he could handle foreign leaders more skillfully than his opponent, a former Secretary of State. He never criticized Putin's moves against Ukraine, his annexation of Crimea, or his support of Bashar al-Assad in Syria. Instead, he brushed past all that with the declaration,

"Wouldn't it be nice if we actually got along with Russia? Wouldn't that be good?"

So when Maggie Haberman and I were preparing on July 20 to conduct our second foreign-policy interview with Trump—the day before he would accept the party's nomination—Russia was high on our list of questions. We stepped into his hotel room in Cleveland just as he was finishing a meeting with Manafort, who shook our hands and quickly stepped out of the room, before any questions might be directed his way.

Trump was distracted and a bit irritated by something he had just heard about himself on television, but he settled in when the questions began, eager to prove himself familiar with every global hot spot. About halfway through the interview, I saw an opening and noted to Trump, "You've been very complimentary of Putin himself."

"No! No, I haven't," he insisted.

SANGER: You said you respected his strength.

TRUMP: He's been complimentary of me. I think Putin and I will get along very well.

We pursued that non sequitur for a while; I was trying to draw him out on why the fact that Putin had been complimentary of the soon-to-be-nominee would in any way affect Trump's judgment about how to deal with an increasingly aggressive adversary. When that went nowhere I tried another route, testing whether he would defend the newest members of NATO.

"I was just in the Baltic States," I told him. "They are seeing submarines off their coasts, they are seeing airplanes they haven't seen since the Cold War coming, bombers doing test runs. If Russia came over the border into Estonia or Latvia, Lithuania, places that Americans don't think about all that often, would you come to their immediate military aid?"

This was, I thought, the bottom-line issue: if Putin wanted Trump to win, it had to be because he thought a Trump victory would undercut the Western allies' confidence that America would defend the alliance. Trump tried to duck:

> **TRUMP:** I don't want to tell you what I'd do because I don't want Putin to know what I'd do. I have a serious chance of becoming president and I'm not like Obama, that every time they send some troops into Iraq or anyplace else, he has a news conference to announce it.

As soon as Maggie and I pressed the point, Trump took refuge in one of his favorite arguments: NATO members are taking us for granted and "aren't paying their bills." So I decided to get a little more specific:

> **SANGER:** My point here is, can the members of NATO, including the new members in the Baltics, count on the United States to come to their military aid if they were attacked by Russia? And count on us fulfilling our obligations—

> **TRUMP:** Have they fulfilled their obligations to us? If they fulfill their obligations to us, the answer is yes.

> **HABERMAN:** And if not?

> **TRUMP:** Well, I'm not saying if not. I'm saying, right now there are many countries that have not fulfilled their obligations to us.

There was our story for the evening before he became the Republican nominee: The first major presidential candidate to cast doubt on whether the United States would come to the defense of treaty allies.

I had one other line of questions I wanted to try: How would he respond to cyberattacks? Particularly those "that are short of war" and "clearly appear to be coming from Russia?"

TRUMP: Well, we're under cyberattack.

SANGER: We're under regular cyberattack. Would you use cyber-weapons before you used military force?

TRUMP: Cyber is absolutely a thing of the future and the present. Look, we're under cyberattack, forget about them. And we don't even know where it's coming from.

SANGER: Some days we do, and some days we don't.

TRUMP: Because we're obsolete. Right now, Russia and China in particular and other places.

SANGER: Would you support the United States not only develop-ing as we are but fielding cyberweapons as an alternative?

TRUMP: Yes. I am a fan of the future, and cyber is the future.

That was as far as we got on how the nominee thought about the newest weapon that Russia and the United States were utilizing in a global strug-gle for power: "Cyber is the future." But worse yet, he fueled our suspicions that at a minimum he was perfectly comfortable with what was clearly Russian interference in the election. And he made us wonder whether, wittingly or unwittingly, he had become Putin's agent of influence.

THE LEAKED EMAILS apparently weren't producing as much news as the GRU-linked hackers had hoped. So the next level of the plan kicked in: activating WikiLeaks.

The first WikiLeaks dump was massive: 44,000 emails, more than 17,000 attachments. And not coincidentally, the deluge started just days after our interview with Trump, and right before the start of the

Democratic National Convention in Philadelphia. The most politically potent of the emails made clear that the DNC leadership was doing whatever it could to make sure Hillary Clinton got the nomination and Bernie Sanders did not.

To anyone watching the nomination process, that was hardly surprising; while the DNC was supposed to be neutral, it was understood in the Democratic leadership that this was Clinton's turn. She had the name recognition and the money and the experience, and many in the party felt she had been denied her chance when Obama came along in 2008. That air of inevitability about her candidacy ended up being one of her greatest liabilities.

Yet the emails that were released in the trove were so blunt and insulting that they played to the divisions within the party, just as Sanders's delegates were showing up in the sweltering heat of Philadelphia. One of the big questions was whether the Russians knew enough by themselves to intensify that division, or whether they had help from Americans who had an interest in undercutting the Democrats.

If the Russian goal was simply to trigger chaos, it worked. Wasserman Schultz, the Florida congresswoman, had to resign as the party's chair just ahead of the convention over which she was set to preside.

And finally the country—or at least anyone following what was happening closely—was waking up. In the midst of the Democratic convention in late July, my colleague Nicole Perlroth and I wrote: "An unusual question is capturing the attention of cyber specialists, Russia experts and Democratic Party leaders in Philadelphia: Is Vladimir V. Putin trying to meddle in the American presidential election?"

Clinton's campaign manager, Robby Mook, accused the Russians of leaking the data "for the purpose of helping Donald Trump," though he cited no evidence.

Mook suggested that Trump's answers to us the week before about whether he would come to NATO's aid marked a watershed moment. Such an allegation seemed unprecedented. Even at the height of the Cold War, we wrote, "it was hard to find a presidential campaign willing to charge that its rival was essentially secretly doing the bidding of a key American adversary." For the first time we raised the question of whether Putin himself was behind the leaks.

That question had already seized the CIA and the NSA. Two days later, in Washington, word began to spread that a preliminary CIA assessment circulating in the White House—deeply classified—concluded with "high confidence" that the Russian government was behind the theft of emails and documents from the Democratic National Committee. It was the first time that the government began to signal that a larger plot was under way.

Yet publicly the White House remained silent. The CIA evidence, my *Times* colleague Eric Schmitt and I wrote, "leaves President Obama and his national security aides with a difficult diplomatic and political decision: whether to publicly accuse the government of President Vladimir V. Putin of engineering the hacking."

In fact, a fight was brewing inside the administration on just that point. What we didn't know at the time was that a disagreement had surfaced among the intelligence agencies. The CIA's "high confidence" was in part based on human sources inside Russia. The NSA was not prepared to sign on; it did not yet have enough signals intelligence and intercepted conversations to say with anything more than "moderate confidence" the hack was a GRU operation, and that Putin had ordered it.

"This went to the heart of Russia's role and intentions," said one senior official who participated in the debate in early August, right after the conventions. "And finally Obama—who is usually pretty cool about these things—got pretty animated. He said, 'I need clarity!' And he didn't have clarity" about who had ordered the hacking, or what its objectives were.

Trump himself seemed to understand what was at stake. "The new joke in town," he wrote on Twitter, "is that Russia leaked the disastrous DNC emails, which should never have been written (stupid), because Putin likes me."

Soon it would not be a joke.

CHAPTER X

THE SLOW AWAKENING

In this new cyber age, we're going to have to make sure that we
continually work to find the right balance of accountability and
openness and transparency that is the hallmark of our democracy.

—*Barack Obama, at his final White House press conference,*
January 18, 2017

N LATE JULY 2016, with emails stolen by GRU-affiliated hackers surfac-
ing every few days on the WikiLeaks website, Victoria Nuland settled
into her office in the State Department. Surrounded by the rugs and
mementos of many foreign trips, she began drawing up her wish list
of measures the US government could take to make life miserable for
Vladimir Putin.

Her list was long, and one of Nuland's colleagues later said "it
veered more toward punishment than deterrence." But as it circulated
among a small group of officials at the State Department and the National Security Council, Nuland's call to action underscored that the
United States had plenty of options.

The list started with the obvious: If Putin wanted to play the game
of releasing embarrassing information, why not let him feel what it was
like to be on the receiving end? (Nuland certainly remembered what
that was like, after the release of her infamous call with the American
ambassador to Ukraine.) American intelligence agencies had assembled

a pretty good picture of Putin's vast financial holdings, spread around the world in secret accounts outside Russia. Many were being held on his behalf by his oligarch friends. Wouldn't there be some justice, Nuland asked her colleagues, in some well-timed revelations about the hundreds of millions, maybe billions, of dollars that Putin had tucked away?

And there was much to reveal about the oligarchs themselves, who had siphoned billions of dollars out of the Russian economy—one reason it was now moribund—and used the money to buy $100 million flats in London. Add to that some revelations about their less-than-savory businesses, and many of those fortunes could be frozen, threatening not only the oligarchs but the lifestyles of their children.

Other options went far beyond embarrassment. Celeste Wallander, the Russia expert at the National Security Council, and Michael Daniel, the cyber-policy chief at the White House who was running the cyber-response team to mitigate the damage, wanted to know what it would take to mount an in-kind response. Daniel, normally a mild-mannered budget specialist, was a veteran of enough Russia hacks to argue that Putin would retreat only if whacked on the nose. Otherwise, Daniel told me later, Putin would "just keep doing what he always does." Was it possible, he and Wallander asked, to fry the servers that DCLeaks and Guccifer 2.0 were using to distribute the stolen emails, or to go directly after WikiLeaks? One idea called for electronic attacks on the GRU, to make it clear that the NSA knew how their command-and-control systems operated and how to screw them up.

But the NSA offered a caution: whatever troubles Washington brought upon the Russians, they argued, would have only temporary impact, and the cost would be huge. The Russians would learn which of their networks had been penetrated by the NSA, and how. "You don't want to do something," said one of the cyber warriors who opposed the ideas, "that has little enduring impact."

One extreme option on the table certainly would have gotten Putin's attention: bringing the Russian economy to a standstill by cutting off its banking system and terminating its connection with SWIFT, the international clearinghouse for banking transactions.

"It was an enormously satisfying response," one of Nuland's colleagues recalled later with a smile, "until we began to think about what it would do to the Europeans," who were still reliant on Russian gas to provide heat through the winter. As one of President Obama's top aides said, "No one was exactly eager to call the Germans and tell them it would be a pretty cold, long winter because the Russians were messing with the Hillary campaign."

By the accounts of three of Obama's top national security aides, none of these recommendations formally made it to President Obama prior to the 2016 election. (Informally, several were discussed with him.) Those advisers at the top of his national-security pyramid—Susan Rice, the national security adviser; Rice's deputy, Avril Haines, who was tasked with leading the "deputies process" to sort out the options; and Lisa Monaco, the homeland security adviser—all argued that while pushing back against the Russians was important, ensuring that the electoral process was secure was their first priority.

"That was our number-one focus," said Denis McDonough, Obama's chief of staff, who agreed with the cautious approach. "The president made it clear that the integrity of the election came first." Making the Russians pay a price was important, but it could wait until the ballots were counted.

So the waiting began.

ONCE THE GRU—through Guccifer 2.0, DCLeaks, and WikiLeaks—began distributing the hacked emails, each revelation of the DNC's infighting or Hillary Clinton's talks at fund-raisers became catnip for political reporters. The content of the leaks overwhelmed the bigger question of whether everyone—starting with the news organizations reporting the contents of the emails—was doing Putin's bidding.

From the time in early August that John Brennan, the CIA director, began sending intelligence reports over to the White House in sealed envelopes, the administration was preoccupied with the possibility that a far larger plot was under way. Perhaps, the officials feared, the DNC hack was only an opening shot, or a distraction. Already reports were trickling in about constant "probes" of election systems in Arizona and Illinois, all traced back to Russian hackers. Was Putin's bigger plan to hack the votes on November 8? And how easy would that be to pull off?

In part, Obama's concern about an Election Day hack grew from the alarm with which a small band of White House aides viewed the contents of Brennan's restricted missives. The envelopes contained reporting from a tiny number of high-ranking Russian informants in Putin's orbit—including at least one source so sensitive that Brennan did not want the reports included in the President's Daily Brief, which had a wide circulation in the White House, the State Department, and the Pentagon. It was largely these sources' accounts of Putin's intentions and orders that made the CIA declare with "high confidence" that the DNC hack was the work of the Russian government at a time when the NSA and other intelligence agencies still harbored doubts. The sources described a coordinated campaign ordered by Putin himself, the ultimate modern-day cyber assault—subtle, deniable, launched on many fronts—incongruously directed from behind the six-hundred-year-old walls of the Kremlin. Putin didn't think Trump could win the election, the CIA concluded. Like just about everyone else, he was betting that Clinton, his nemesis, would prevail. But he was hoping to weaken her by fueling a post-election-day narrative that she had stolen the election by vote tampering.

Brennan later argued that Putin and his top aides had two goals: "Their first objective was to undermine the credibility and integrity of the US electoral process. They were trying to damage Hillary Clinton. They thought she would be elected, and they wanted her bloodied by the time she was going to be inaugurated," he said in a conversation in Aspen, Colorado, in the summer of 2017, six months after he had left the CIA. But Putin was hedging his bets, Brennan surmised, by "also trying to promote the prospects of Mr. Trump."

Whether Russia could succeed in tampering with votes depended on the resilience of the voting infrastructure. That infrastructure was run by state officials, who were reluctant to let the federal government become involved too closely. Some holes in the system in a few of the most critical swing states were well known: Pennsylvania, notably, had almost no paper backup for its voting machines. Even if a post-election audit of the vote were conducted, there was no viable way to confirm that votes were reported the way they had actually been cast. Other states had similar vulnerabilities. But no one had a detailed, nationwide picture of the problem. Initially, "no one really understood what the vulnerability of the election system was—whether you could hack the vote," Avril Haines told me later.

To understand the vulnerabilities, Obama secretly ordered a National Intelligence Estimate, a usually classified document prepared by an independent group called the National Intelligence Council that looks at big, complex topics and gives detailed assessments of American vulnerabilities and capabilities. The NIC was known for its independence, and occasionally for its willingness to challenge conventional wisdom. Past NIEs had examined Iran's nuclear capability, the stability of the Chinese leadership, and even the national security implications of climate change. But never before had the group been asked to take a comprehensive look at the susceptibility of the American election system to outside influence.

While the administration waited for the report, Trump began warning about election-machine tampering, seemingly laying a foundation for the argument, to be made on November 9, that Hillary Clinton had won fraudulently. He began hitting this theme in the friendly environment of the Sean Hannity show on Fox News on August 1, and ramped it up at a rally, where a standard campaign line emerged: "I'm afraid the election is going to be rigged." He never talked about evidence, or who would do the rigging. He didn't have to: the rhetoric tapped into his core supporters' belief that somehow the "deep state" was going to manipulate events to deny him the presidency. He'd later tell a rally in Wisconsin: "Remember, we are competing in a rigged election. . . . They even want to try and rig the election at the polling booths, where so many cities are corrupt and voter fraud is all too common."

It all fit a disturbing pattern. The publication of the DNC material seemed highly coordinated. Russian propaganda was in overdrive; while no one yet understood the extent of the problem, there were reports of fictitious news stories about Clinton's health, which usually were stuck in an echo chamber, bouncing between the Russian TV network RT and Breitbart News, Steve Bannon's mouthpiece. "I didn't realize at the time that two-thirds of American adults get their news through social media," said Haines, who was among the most thoughtful members of Obama's team about the impact of social movements on democratic processes. "So while we knew something about Russian efforts to manipulate social media, I think it is fair to say that we did not recognize the extent of the vulnerability."

Obama's vacation on Martha's Vineyard created an informal deadline for his national-security team. By the time he returned to Washington in the last week in August, they knew he would be looking for options, starting with how to protect the electoral infrastructure.

Jeh Johnson, the former Defense Department general counsel who was by then the secretary of homeland security, began making the case, in private and in public, that America's election system was "critical infrastructure" and deserved special protection—the way the power grid did, or the Lincoln Memorial. It seemed a convincing argument: if the undergirding of American democracy, its ability to conduct free and fair elections, didn't constitute "critical infrastructure," what would? But when Johnson arranged a conference call with state election officials around the country, the disconnect was obvious. He described "troubling reports" he had on his desk of probing and scanning of the election systems in Arizona and other states. But he encountered a wall of suspicion; if he hoped to get support for a federal emergency

initiative to help the state election boards address their cyber vulnerabilities, he was badly mistaken.

"Let's just say I didn't hear much enthusiasm," he told me when I
went to see him with my colleague Charlie Savage at the old naval base
that serves as the department's headquarters and emergency operations
center. The secretary of state of Georgia, Brian Kemp, told Johnson
that he was certain the so-called evidence of hacking was a pretext
for the federal government to try to take over the state-run election
systems. (Kemp later accused the Department of Homeland Security
of hacking into his state's systems to scan them for vulnerabilities and
left the impression that it was Washington, not Moscow, that most
worried him.)

During our interview, Johnson never uttered the word "Russia,"
even though we all knew who was responsible for the effort to break
into the voter registration lists. He was still forbidden, at that moment,
from saying the obvious, because the obvious was still treated as classified information. Yet Johnson's evidence was mounting. In June, Arizona officials discovered that the passwords that belonged to an election
official had been stolen, and they feared that a hacker using them could
get inside the registration system. They took the registration database
offline for ten days to conduct a forensic analysis of whether data had
been changed. In Illinois, there was a deeper panic: the registration
system was pierced and voter information siphoned off. The forensics
suggested the hack was engineered by known Russian groups. Inside
Johnson's homeland-security headquarters, the cyber teams worried
that once hackers got into a registration system, they could change
Social Security numbers or delete voters from the rolls.

"That's all it would have taken to create chaos on Election Day," one
senior White House official told me. "You didn't have to change much."
Few said so at the time, but months after the election Homeland Security said it had seen evidence of similar probes into the systems of
roughly three dozen other states. No one would say why they did not
reveal that information at the time.

The fears, while rampant, were still based on conjecture: Russian hackers had essentially been caught scouting the systems, but not changing anything. And because none of the state officials had security clearances, Johnson's phone call, from a vacation spot in the Adirondacks, was a failure. He had been prohibited from providing the state officials with any specifics. The classification rules—presumably intended to keep the Russians from learning that their activities were being watched—impeded Johnson's ability to make his case. Once again, the reflexive assumption that all evidence of cyberattacks had to be kept highly classified cost America dearly.

To make matters worse, Johnson never detailed the evidence that overwhelmingly suggested Russia was behind the probes into the voting system. A written FBI warning to the states said only that information had been "exfiltrated" from Arizona's system, but it did not indicate where that information was headed. Because the doubts of other intelligence agencies had not yet been resolved, the official position of the US government was to make no accusations about who was behind the hacking. "It was the worst, most vague briefing I've ever heard a government official conduct," one official said of the call. "It wasn't Jeh's fault—he was following the rules. But he could provide no evidence." James Clapper ran into a similar problem: he had seen all the evidence but told me that summer he could "make no calls on attribution" until the disparate assessments of the intelligence agencies came into line. His caution was understandable. But it was also costly: the intelligence agency's paranoia about protecting sources and methods got in the way of warning the targets of the hacks—the election commissions in fifty states—that one of the world's most cyber-savvy nations had them in its sights.

Brennan, meanwhile, had quietly assembled a task force of CIA, NSA, and FBI experts to sort through the evidence. And as his sense of alarm increased, he decided that he needed to personally brief the Senate and House leadership about the Russian infiltrations. It was not an easy task: Most of the leaders were scattered around the country, away from secure phones. One by one he got to them; they had security clearances, so he could paint a picture of Russia's efforts with details that Johnson was forbidden from mentioning.

After Harry Reid, the Senate Democratic leader, received his briefing via a secure phone in Las Vegas, he was agitated and fearful that the government was under-responding to the threat. Perhaps because I had been writing on the subject through the summer, he called me in Vermont, where I was failing in an effort to take a last few days of vacation before the election entered its final phase. He had just received a lengthy briefing from a "senior intelligence official," he told me; there was little doubt in my mind that it was Brennan, since he had been so fixated on the Russia issue in recent weeks. Reid would not offer the details of what he had been told, because they were classified, to his obvious frustration. But he did provide his takeaway: "Putin is trying to steal this election," he told me. Ever the vote counter, he argued that if Russia concentrated on "less than six" swing states, it could alter the outcome.

CLEARLY, VLADIMIR PUTIN would have to be confronted about the evidence from the DNC and the probes on the state election systems. The debate was over how to do it.

Obama's first rule of foreign policy, described to my colleague Mark Landler and others on *Air Force One* during a trip to Asia, was straightforward: "Don't do stupid shit." (He made the reporters repeat it in unison.) As a caution, it wasn't bad; a lot of the worst moves in American foreign policy in the previous two decades had begun with stupid-shit decisions. But as a principle for dealing with Vladimir Putin, it didn't offer much detailed guidance. Antony Blinken, the deputy secretary of state, put it succinctly: Since no one really understood if the Russians had planted code in the election systems—a booby trap that could be triggered on November 8—the cautious approach was to proceed slowly. "You never want to start a contest like this unless you have a reasonable assessment of where it will end up," Blinken told me. Brennan voiced the concern only slightly differently: No one wanted "an escalatory cycle in the middle of a presidential campaign."

Obama was particularly concerned about appearing partisan—or, by making public declarations about Russia's actions, playing into Putin's hands by conceding, before a single ballot was cast, that the election had been compromised. So the White House developed a two-part plan: get the leaders of Congress, Democrats and Republicans, to issue a joint statement condemning Russia's actions, and then have Obama confront Putin at a summit meeting they were both planning to attend in early September.

Obama dispatched Lisa Monaco, along with James Comey, the FBI director, and Jeh Johnson, to Capitol Hill to explain how the federal government was prepared to help the states.

As soon as they got into the session with twelve congressional leaders, led by Mitch McConnell, it went bad. "It devolved into a partisan debate," Monaco later told me. "McConnell simply disbelieved what we were telling him." He chastised the intelligence officials for buying into what he claimed was Obama administration spin, recalled one of the other senators present. Comey tried to make the point that Russia had engaged in this kind of activity before, but this time it was on a far broader scale. The argument made no difference. It became clear that McConnell would not sign on to any statement blaming the Russians.

"It was one of the most dispiriting days I ever had in government," Monaco concluded. A subsequent, smaller session that Obama held in the Oval Office did not end much better.

Obama's summit meeting with Putin, on September 5, was planned as the showdown. As they entered the ninety-minute session in Hangzhou, there were none of the forced pleasantries that usually begin such sessions: Knowing cameras were trained on them, they stared each other down like two sumo wrestlers waiting for the signal to begin combat. Then they headed into a one-on-one discussion. The accounts of how strongly Obama threatened Putin depend on who was telling the story. But his essential warning was that the United States had the power to destroy the Russian economy by cutting off its transactions— and would use that power if American officials believed Russia interceded in the election.

Obama emerged from the session wondering aloud whether Putin was content to live with a "constant, low-grade conflict." He was specifically referring to Ukraine, but he could have been talking about any of the arenas in which Putin relished his role as a great disrupter. It seemed clear that to Putin, constant, low-grade conflict was just fine; it was the only affordable way to restore Russia's eminence on the global stage. "It shouldn't come as a big shock to people," James Clapper, the rare Cold War veteran in Obama's top ranks, said after the Putin meeting. "I think it's more dramatic maybe because now they have the cyber tools."

The administration continued to envelop its debates in great secrecy. The video feeds of meetings at the National Security Council were shut off, much as they were during the run-up to the bin Laden raid. Susan Rice kept tight control of who knew the meetings were happening; always worried about leaks, she feared in this case that they would force Obama's hand.

Only long after the election was over were officials willing to explain the full reason for the switched-off video and the secrecy. In fact, the president's top advisers had received a detailed plan from the National Security Agency and Cyber Command about possible retaliatory strikes against Russia. Some would have fried the servers used to mount the Russian attacks against US targets; others would have put the Internet Research Agency out of action; still more were designed to embarrass Putin or make his money disappear. "It was strikingly detailed," one former official said.

The ideas were limited to a handful of top officials: many of the White House and State Department senior officials working on Russia were not "read in" to the details. But again, Obama's top aides hesitated. They had begun to see some evidence that the Russians were backing off; the probing of the state election systems had slowed dramatically after the Obama–Putin encounter. Hitting the Russians at that moment, just when it looked like they may have gotten the message, seemed counterproductive.

Around the same time, the results of the National Intelligence Estimate about the vulnerability of the election system began to circulate. In a rare bit of good news, the National Intelligence Council concluded that hacking the election machines themselves on a broad scale, while not impossible, would be a daunting job. Most voting machines were offline, meaning that hackers would need a physical presence in key polling places to interfere with the results. Theoretically, it was possible to get inside the software that was downloaded into the machines in advance of an election, but since every locality had a different ballot, and often a mix of voting hardware, it would be a complex operation to pull off. At the White House, the staff was clearly relieved.

At least until Clapper spoke up. He warned that if the Russians truly wanted to escalate, they had another easy path: their implants were already deep inside the American electric grid. Forget hacking the voting machines; the most efficient way to turn Election Day into a chaotic, finger-pointing mess would be to plunge key cities into darkness, even for just a few hours.

There was "a sort of silence for a moment," one participant in that meeting recalled, "and you could sense that people were just letting that sink in."

SOMETHING ELSE HAD sunk in at Fort Meade: Not only were the Russians inside the election infrastructure; they might well be inside the Tailored Access Operations unit, the operations center for America's cyber wars.

In mid-August, when the Democrats were still struggling to figure out what the Russian hackers were doing to them, the NSA discovered that it wasn't only campaign memos that were suddenly showing up on the Internet. So were samples of the tools the TAO had used to break into the computer networks of Russia, China, and Iran, among others.

The tools—everything from code designed to exploit vulnerabilities in Microsoft systems to actual instruction manuals for conducting cyberattacks—were being posted by a group that called itself the Shadow Brokers. The agency's cyber warriors knew that the code being posted was malware they had written. It was the code that allowed the NSA to place implants in foreign systems, where they could lurk unseen for years—unless the target knew what the malware looked like. And the Shadow Brokers were offering a product catalog.

Inside the NSA, this breach was regarded as a far greater debacle than the Snowden affair. For all the publicity and media attention around Snowden, a dark if compelling character who could still command headlines from his exile in Russia, the Shadow Brokers were inflicting far more damage. Snowden released code words and PowerPoints describing what amounted to battle plans. The Shadow Brokers had their hands on actual code, the weapons themselves. These had cost tens of millions of dollars to create, implant, and exploit. Now they were posted for all to see—and for every other cyber player, from North Korea to Iran, to turn to their own uses.

"People were stunned," one former employee of the TAO said. "It was like working at Coca-Cola, and waking up to discover that someone had just put the secret formula on the Internet."

The initial dump was followed by many more, wrapped in taunts, broken English, a good deal of profanity, and a lot of references to the chaos of American politics. The Shadow Brokers were promising a "monthly dump service" of stolen tools and leaving hints—perhaps misdirection—that Russian hackers were behind it all. "Russian security peoples," one missive read, "is becoming Russian hackers at nights, but only full moons."

The posts raised many questions. Was this the work of the Russians, and if so was it the GRU, trolling the NSA the way it was trolling the Democrats? Did the GRU's hackers break into the TAO's digital safe—which seemed unlikely—or did they turn an insider, maybe several? And was this hack related to another loss of cyber tools, equally embarrassing, from the CIA's Center for Cyber Intelligence, which had been appearing for several months on the WikiLeaks site under the name "Vault 7"?

Most important, was there an implicit message in the publication of these tools—a threat that if Obama came after the Russians too hard for the election hack, more of the NSA's code would become public?

Inside the NSA, these questions were rampant. But they were never uttered in public. The NSA's counterintelligence investigators, called the Q Group, went on a broad hunt for "undiscovered Snowdens," as one senior official put it. The agency, which had been forced to open up a bit after Snowden, explaining its missions and the legal basis for where it would and would not spy, shut the gates again. Suddenly employees found themselves subjected to polygraph tests, and some were suspended from their jobs. Some departed; a hotshot hacker for the TAO might command upwards of $80,000 a year from the NSA but could make multiples of that figure in the private sector. Many had been willing to make less money to break into foreign systems to defend American interests. But now they reconsidered: was it worth giving up the extra income if you were treated with suspicion at work and strapped to a lie-detector machine?

"Snowden killed morale," one TAO analyst told us when Scott Shane, Nicole Perlroth, and I dug into the tale of the Shadow Brokers. "But at least we knew who he was. Now you have a situation where the agency is questioning people who have been 100 percent mission-oriented, telling them they are liars."

The worst part was the fear that came from not knowing if the hemorrhaging had stopped. With their implants in foreign systems exposed, the NSA temporarily went dark. At a moment when the White House and the Pentagon were demanding more options on Russia and a stepped-up campaign against ISIS, the agency was busy building new tools because the old ones had been blown.

Adm. Rogers and other leaders at the agency strongly suspected the Russians were either behind the attack or the beneficiaries of it. The NSA had already been stung by Moscow in 2015—twice. First, Kaspersky Lab, Russia's most famous cybersecurity group and a maker of antivirus software, had published a report about the activities of what it called "The Equation Group," detailing malware implanted in dozens of countries. You didn't have to read between the lines very carefully to see that the Equation Group was really the Tailored Access Operations unit; some of the malware that Kaspersky highlighted as the group's handiwork included code from the Olympic Games attacks on Iran. Then, to provoke the NSA further, Kaspersky issued new versions of its antivirus software, used by 400 million people around the world, that detected some of the TAO's malware and neutralized it.

Then the Shadow Brokers began crowing. "We hack Equation Group," they wrote. "We find many many Equation Group cyber weapons."

It was not clear they did "hack" the Equation Group. But there were two incidents involving NSA contractors that likely seem linked to how the TAO's darkest secrets got out—and, most officials believe, ended up in the hands of the Russians.

The first occurred in late 2014 or early 2015, when a sixty-seven-year-old NSA employee, Nghia H. Pho, took home classified documents. Pho, a native of Vietnam who became a naturalized American citizen, worked deep inside the TAO for a decade, starting in 2006. But after about four years on the job, according to court documents, he started to bring home classified documents, many of them in digital form.

It turned out that Pho was using Kaspersky's antivirus software, which someone, likely the Russian intelligence agencies, had brilliantly manipulated to search for NSA code words—and Pho had apparently brought home documents that contained some of those words. In effect, Kaspersky's antivirus products appeared to be giving Russian intelligence a back door into any computer it was installed on.

For Rogers, the Pho case was a disaster; he had been brought in to clean up after the Snowden affair, not let other vulnerabilities fester. Then in early October 2016, things got even worse. Investigators trying to crack the Shadow Brokers case arrested another Booz Allen Hamilton contractor, Harold Martin III, whose house and car in a suburban tract in Glen Burnie, Maryland, were brimming over with classified documents, many from the TAO. Martin kept much of his stolen trove electronically, and the data amounted to "many terabytes" of information, according to the FBI. And it wasn't limited to the NSA: court filings said he stole material, during previous posts, from the CIA, Cyber Command, and the Pentagon.

It did not look as if Martin was working for the Russians, but the material in his possession included some of the Tailored Access Operations unit's tools that were ultimately put up for sale by Shadow Brokers. However, just because Martin had the materials, it did not necessarily mean that Shadow Brokers had acquired them from him, which left open the possibility that there were even further leaks from the NSA's systems.

Rogers was now under more pressure than ever. The Pentagon was hammering him about where all the leaks were coming from. He was issued a reprimand, officials say, though the NSA would not discuss the subject. Nor would Rogers. And the timing was awful: he was under fire at the same moment the White House was asking for cyber options to deal with Russia, something that could deter Putin from further action against the United States.

But what if Putin's hackers already had pieces of the NSA's arsenal?

THE SHADOW BROKERS disclosures alarmed the intelligence agencies because they suggested Snowden was not a one-off affair; the nation's cyber warriors had been repeatedly, deeply compromised. But Obama and his team did not have time to deal with that issue. The debate inside the Situation Room was about how to handle Russia, and it bore no resemblance to the debate out on the campaign trail.

There, Trump did everything he could to cast doubt on the reliability of the intelligence about Russian interference. It was "impossible" to trace cyberattacks, he said, which was patently untrue. Hard, yes. Impossible, no. At his first debate with Hillary Clinton, he argued there was no evidence that Russia was responsible, and famously added that it could have been the Chinese or "somebody sitting on their bed that weighs four hundred pounds." Ridiculous as it sounded, it was a reminder of how in the public's mind, cyberattacks seemed so complex and mysterious that the topic lent itself to false claims and political misdirection.

Trump's contentions stepped up the pressure on the administration to name the Russians—and to provide some evidence. It wasn't clear either would happen. Obama had pulled back from naming them after the network intrusions at the White House and the State Department, and a later, bold intrusion in 2015 into the computers of the Joint Chiefs of Staff—it was entirely possible his overcaution would again prevail. Yet by October, Obama had concluded that the campaign attacks were different. They were not just espionage; they constituted an attack on American values and institutions—and were more akin to the Sony hacks, which he viewed as an attack on free expression. Obama was hesitant to come to the podium and call out the Russians. It would look too political, he told his aides.

So on October 7, Clapper released a statement from the Office of the Director of National Intelligence, also signed by Jeh Johnson, who was still trying to convince state election boards to let the federal government scan their systems for signs of malware. (Curiously, Comey, fearful of sinking the FBI further into the political campaign, declined to sign the warning. Three weeks later he plunged right into that maelstrom by reopening, then reclosing, the investigation into Hillary Clinton's emails.) The statement confirmed what the country already knew—at least, those who had been paying attention or hadn't dismissed the intelligence as politically driven: "The US intelligence community is confident that the Russian government directed the recent compromises of emails from US persons and institutions, including from US political organizations." The statement also said that "some states have also seen scanning and probing of their election-related systems" from Russia, though it stopped short of accusing the Russian government.

Inside the White House, there had been a vigorous debate about whether to accuse Putin directly, and whether, if he was named, it would provoke him to further action. In the end, the statement was watered down: "We believe, based on the scope and sensitivity of these efforts, that only Russia's senior-most officials could have authorized these activities." To avoid public panic—and the easy win that would give Putin—the statement included the carefully hedged assessment that "it would be extremely difficult for someone, including a nation-state actor, to alter actual ballot counts or election results by cyber attack or intrusion."

The statement was only three paragraphs. It offered not a shred of evidence, even though there was plenty, an omission that played into Trump's hands because without proof, Trump could continue to claim Russia may have had nothing to do with it. This was the first time in history the United States had accused a foreign power of seeking to manipulate a presidential election on a broad scale.

It would have been huge news, save for the spectacularly bad timing.

Just as the government's statement about Russian interference was beginning to circulate, the *Washington Post* published news of the *Access Hollywood* audiotape of Trump, in 2005, describing how "when you're a star, they let you do it, you can do anything," including actions that clearly would constitute sexual assault. It seemed for a moment that the one-two punch—the Russians' interfering on his behalf and the surreptitious tape—would finish him off.

But as Clinton herself wrote, the events proved "the old Washington cliché about how the 'drip, drip' of scandal can be even more damaging over time than a single, really bad story. Trump's tape was like a bomb going off, and the damage was immediate and severe. But no other tapes emerged, so there was nowhere else for the story to go." The tape and its aftermath largely obliterated much discussion of the intelligence

findings. Within an hour, WikiLeaks began publishing John Podesta's emails, which had been stolen back in March. Suddenly the focus
changed to what Hillary had said in speeches to Goldman Sachs, and
the internal conversations about her shortcomings as a candidate. Putin
had caught a lucky break, again. The Podesta emails dominated the air-
waves in the last month of the campaign, but how they became public
did not. And Obama decided that any sanctions against Russia should
proceed only if it appeared that his warning to Putin—reiterated in a
secret letter to the Russian leader that former members of the administration will not discuss—had done no good.

The FBI and Brennan reported a continued decrease in Russian
"probes" of the state election systems. No one knew exactly how to
interpret that fact—it was possible the Russians already had their implants in the systems they had targeted. But as one senior aide said, "it
wouldn't have made sense to begin sanctions" just when the Russians
were backing away.

The administration decided to put off the question of deterrence—
or punishment—for a few weeks, until after the election.

ELECTION DAY CAME and went with no penalties for Putin, almost no
evidence of suspicious cyber activity at the polls, and the election of a
candidate who said the hacking probably never happened and probably wasn't the Russians if it did. Suddenly all the decisions that had
been made about pushing back against the Russians required reexami-
nation. "There had been an assumption that Hillary would win and
we'd have time to figure out a set of actions that could be carried into
the next administration," one senior official said. "Suddenly we had to
come up with some steps that couldn't be reversed."

Obama's team was stunned. John Kerry pushed for a September 11–style commission to set out the facts of the Russian intrusion;
that idea got shot down. So did new versions of Victoria Nuland's proposals about releasing embarrassing information about Putin himself.

Even then, in the wake of Trump's astonishing win, Obama still could not bring himself to take immediate, strong action against Putin, the oligarchs, or the GRU. He worried that the United States would lose the moral high ground. But you could hear the regret in Obama's voice when he talked to reporters in mid-December. He laid out the facts, including the telling admission that he didn't know about the hacking of the DNC until "the beginning of the summer" of 2016, without mentioning that that was nine months after the FBI made the first call to DNC headquarters. "My hope is that the president-elect is going to similarly be concerned with making sure that we don't have potential foreign influence in our election process. I don't think any American wants that. And that shouldn't be a source of argument."

Of course, it became just that. Obama seemed determined not to place the hack into the context of the far larger plan Putin was executing, one whose shadow had lengthened across his second term: the attacks on Ukraine, the intrusions into the American electrical grid, the digital battle with Russian hackers for control over the unclassified network in Obama's own White House. "This was not some elaborate, complicated espionage scheme," he said, dismissing the emails the Russians had released as "pretty routine stuff, some of it embarrassing or uncomfortable." The big concern, he suggested, was the way everyone—the media, the voters—fixated on it.

And he defended his decision to stay quiet until then. "My principal goal leading up to the election was making sure that the election itself went off without a hitch, that it was not tarnished, and that it did not feed any sense in the public that somehow tampering had taken place with the actual process of voting." Now, he said, "that does not mean that we are not going to respond."

When the response came, it was right out of the diplomatic playbook. Thirty-five Russian "diplomats" were thrown out of the country, most of them spies, some suspected of abetting the hacking into American infrastructure. A few Russian facilities were closed, including the consulate in San Francisco, where black wisps of smoke rose from the chimney as the Russians burned paperwork. The White House also announced the closure of two Russian diplomatic properties, in Long Island and Maryland. What the administration did not say was that one of them was being used by the Russians to bore underground and tap into a major telephone trunk line that would presumably give them access to both phone conversations and electronic messaging—and perhaps another pathway into American computer networks. But over-all it was, as one of Obama's own aides said, "the perfect nineteenth-century response to a twenty-first-century problem."

As a secret parting shot, Obama ordered that some code—easy to discover—be placed in Russian systems, a "Kilroy was here" message that was later spun by some as a time bomb left in Russian networks. If so, it was never armed. As a deterrent, it wasn't much of a success. In fact, the Russians had largely won. As Michael Hayden, the former CIA and NSA director, said, it was "the most successful covert opera-tion in history."

THE OBAMA ADMINISTRATION's parting cyber sanctions triggered the first scandal of the Trump transition: Trump's national-security desig-nate, Lt. Gen. Michael Flynn, quietly told the Russian ambassador he would look at the sanctions as soon as Trump took office. He later lied about the conversation, got caught, resigned, and pled guilty to lying to the FBI.

Meanwhile, in a bizarre briefing at Trump Tower conducted by Clapper, Brennan, Comey, and Rogers, Trump was presented with the classified evidence of Putin's role in hacking the election. He later dis-missed Clapper and Brennan, career intelligence officers, as "political hacks." He fired Comey, largely for pursuing the Russia investigation and refusing to declare his loyalty to the new president.

Comey's firing resulted in the appointment of a special counsel, Robert Mueller, who peeled away layer after layer of evidence about the Trump campaign's involvement with the Russians. Then Comey went in front of Congress and said that not only had the Russians intervened, but they would try to do it again. "It's not a Republican thing or a Democratic thing . . . ," he said. "They're going to come for whatever party they choose to try and work on behalf of. And they're not devoted to either, in my experience. They're just about their own advantage. And they will be back."

Trump, his own advisers conceded to me, refused to discuss the Russia hack: He viewed the entire investigation as an effort to undercut his legitimacy. As a result, he seemed unable to design a strategy for dealing with Moscow—which put Putin in the driver's seat.

It wasn't until July 7, 2017, six months into his presidency, that Donald Trump finally met Vladimir Putin. They had circled each other for years, each thinking about how to manipulate the other, before they sat down in Hamburg, Germany, at the edge of a Group of 20 summit meeting,

By that time it was becoming clear to Putin that his bet that Trump would eradicate the sanctions choking Russia's economy was failing, spectacularly. Even the Republican Congress, usually loyal to Trump to a fault, was on the verge of passing new sanctions against Russia for its election interference. Trump could not veto it. And the arms race was accelerating.

Putin and Trump talked behind closed doors for two and a quarter hours. Trump brought only Rex Tillerson, the beleaguered secretary of state, who would be unceremoniously fired over Twitter eight months later. By Tillerson's account to a group of us after the meeting, they covered everything from Syria to the future of Ukraine. But Tillerson also said that "they had a very robust and lengthy exchange" on the election hack and agreed to a meeting of American and Russian officials to create "a framework in which we have some capability to judge what is happening in the cyber world and who to hold accountable."

Shortly after the summit broke up, Trump headed for *Air Force One,* and called me once he was in the air. He wanted to describe his first encounter with Putin. Most of Trump's call was off the record, but some of what he told me that afternoon he repeated at various times in the next few days, in talking about the session.

He had raised the election hacking three times, he said, and Putin had denied involvement each time. But what was more remarkable was the explanation he gave. Trump had asked if he was involved in election meddling, he said. Putin denied it, and said, "If we did, we wouldn't have gotten caught, because we're professionals."

Trump told me he believed that explanation. "I thought that was a good point because they are some of the best in the world," he said, a line he repeated nearly verbatim two days later. I asked Trump whether he believed Putin's denial despite the evidence that Clapper, Brennan, Comey, and Rogers had shown him six months before—some of it drawn from intercepts of Russian communications. Clapper and Brennan, Trump responded, were two of the "most political" intelligence people he knew, and Comey was "a leaker."

He clearly considered the Russia-and-the-election issue closed. Then our line got cut off as *Air Force One* ascended.

MORE THAN TWO years later, with the benefit of hindsight, the sequence of missed signals and misjudgments that allowed Russia to interfere in an American election seems incomprehensible and unforgivable—and yet completely predictable for a nation that did not fully comprehend the many varieties of cyber conflict.

Many of the initial mistakes were born of bureaucratic inertia and lack of imagination: The FBI fumbled the investigation, and the DNC's staff was asleep at the wheel. That deadly combination allowed the Russian hackers complete freedom to rummage through the DNC's files before the party's leadership and the president of the United States were briefed about what was happening. The lost time proved disastrous.

If the Russians had struck at our election system in some more obvious way—poisoning candidates it opposed, for example, as it has poisoned dissidents—any president would have called them out and responded. Only because the gray zone of cyber conflict gave the Russians cover did Obama hesitate. By the time he responded, after the election, it was too late.

We are likely to pay for that failure for years to come. As James Comey said of the Russians: "They will be back."

Some who look back now on the decisions made in the summer and fall of 2016—politicians and national-security staffers, intelligence officials and FBI agents, Russia specialists and journalists—have described the sequence of events as a massive intelligence failure. But it wasn't a failure in the classic sense of the phrase. It didn't launch America into a war on false pretenses, and it didn't underestimate the progress of Russia's nuclear program, or North Korea's for that matter. Rather, it was a failure to confront how skillfully and creatively the Russians were using newly minted cyber skills around the world and how potent a weapon they could become in widening America's political and social fault lines. In our fixation on the types of cyberattacks we thought we understood—against power grids or banks or nuclear centrifuges—we missed the turn toward manipulating voters.

"Was this the cyber 9/11?" Susan Gordon, the deputy director of national intelligence, and a longtime CIA analyst, asked a year after the election. "I don't know. Maybe it was, because it affected something far more fundamental than our electric grid—it affected the workings of our democracy. But it was hard to know that at the time."

Still, it didn't look much like a 9/11 attack. That was designed to be a spectacular, singular event. The undercutting of the American election system was the opposite. It stretched over many months. Initially, it was hard to detect—and once detected, it was designed to be a deniable operation. Part of it was discovered before Americans went to the polls, but the social-media campaign became evident only months after Donald Trump was elected. And to this day, no one can prove whether it actually affected the election's outcome—in fact, the dispute over its effects widened the political divide, as the Russians intended.

It also took advantage of a technological perfect storm. Just as the Russian effort was ramping up, companies like Facebook were making changes that played right into Moscow's hands. Facebook's conscious transition to becoming one of the world's leading global news delivery systems, and tailoring that news to the tastes of each recipient, meshed beautifully with Russia's desire to accentuate the divisions in American society. Worse yet, Facebook (and Twitter) put far too little energy into understanding how their systems were being hijacked by young Russian trolls and bot makers who knew how to take advantage of the algorithms that made the systems work. It is impossible to know whether the Russian campaign succeeded in changing hearts and minds. Yet the truth remains that the tech firms who were so repelled by Donald Trump invented a system that may have helped elect him.

Without question, the Russian decision to move from an espionage operation aimed at disrupting the election to an effort to put Donald Trump in office propelled the country into an entirely new place. We now think about the effects of cyberattacks entirely differently. Just five years before, our worry was China's theft of intellectual property. Then came North Korea's efforts at revenge, and Iran's threats to the financial system.

But the Russian attacks exposed more than the Obama administration's lack of a playbook for cyber conflict, despite years of ever-escalating, ever-more-ingenious attacks. Russia's multifaceted, Gerasimov-inspired approach underscored the administration's failure to anticipate that cyberattacks can be used to undermine more than banks, databases, and electrical grids—they can be used to fray the civic threads that hold together democracy itself.

CHAPTER XI

THREE CRISES IN THE VALLEY

If you had asked me, when I got started with Facebook, if one of the central things I'd need to work on now is preventing governments from interfering in each other's elections, there's no way I thought that's what I'd be doing, if we talked in 2004 in my dorm room.

—*Mark Zuckerberg, on the use and abuse of Facebook data in the presidential elections, March 2018*

THE SUICIDE BOMB went off outside the Stade de France in Saint-Denis at 9:20 p.m. on November 13, 2015, the first of three. Nine minutes later, shootings began on the streets of Paris, triggering panic among diners, who raced to the backs of restaurants, trying to avoid becoming the latest victims of ISIS.

Twenty minutes after the first attack, the shooters entered the nearby Bataclan music hall as the song "Kiss the Devil" poured through its speakers. A few people in the back heard the shout of *"Allahu Akbar."* And the gunfire broke out. First from the mezzanine level, then from the aisles, as the shooters walked up and down, firing at anyone still moving. By the time the siege of Paris ended, a little after midnight, 130 people had died, two-thirds of them trapped in the killing zone of the theater.

What happened next was predictable: the claim of responsibility by ISIS, the recriminations over who let jihadists move freely across the passport-free borders of France and Belgium, President François Hollande's vow to "be unforgiving with the barbarians from Daesh." But then came a fascinating, and very quiet, battlefield partnership between Facebook, the FBI, and the French authorities to hunt down the rest of the ISIS cell.

As the bodies of some of the nine terrorists were photographed, the police turned to Facebook for help in identifying them and their friends around Europe and the world. They were looking for ISIS members who had helped prepare for the attacks or who were readying future attacks. It quickly became clear that several of the terrorists had multiple Facebook accounts that reflected their split lives. Some showed normal European lifestyles, while others, under noms de guerre, portrayed lives of struggle against the West. The French and the FBI obtained court orders within minutes or hours, issued by judges in New York standing by to help. Facebook could then legally share data on the suspected terrorists. It turned over a treasure trove of links between the accounts and specific cell-phone numbers. In some cases the police even had IP addresses from the last places the terrorists had signed into their accounts.

"Once we had a cell-phone number," said one of the people involved in the investigation, "the game was over."

French and Belgian police, with the help of European intelligence agencies and the National Counterterrorism Center back in Virginia, began to triangulate where the attackers were holed up. By November 15, the police were raiding hundreds of locations. A few days later they engaged in a gun battle with ISIS in Saint-Denis. More raids in Belgium followed.

At first glance, the speed and success of the manhunt provided vivid evidence of how social media, when smartly harvested, could be turned against the terror organizations that used the same tools to recruit, organize, and communicate. The connections drawn so quickly from the Facebook community of ISIS supporters helped to dismantle the cell's support structure. No one will ever know how many lives that swift action saved.

But, of course, the lesson of the Paris attack is more complicated. The only issue that animated the Europeans more than hunting down ISIS members in their midst was the competing instinct to protect the privacy of their citizens from Internet behemoths like Facebook. Not long after the attack, Facebook executives met with EU officials about new rules going into effect to protect the privacy of European citizens—rules that limited the kind of information that social-media companies and cloud-storage providers can retain. When the executives reviewed the list of what they could no longer keep, they warned the EU representatives that this was just the kind of data—phone numbers and IP addresses—that had enabled them to help the police track down the Paris attackers. If they could not retain it, they could not help when the next attack happened.

"They didn't care," one of the Facebook executives told me. "They said that's a problem for the intelligence agencies, not the regulators. And the two clearly weren't talking to each other."

IF THERE IS one lesson that emerged from years of trying to find, follow, and disrupt terrorists, it is that the same countries that figured out how to destroy centrifuges from afar and disrupt power grids and missile systems were stymied by how to deal with what has come to be called "weaponized social media." The term itself is a subject of debate. Is a recruiting message that contains a call-to-arms a weaponization of social media, or is it merely what used to be called propaganda, just distributed more quickly and widely? What about a beheading video meant to instill fear, or far more subtle messages of the kind Putin used to widen social and religious divides?

Given the billions of dollars that governments spend to build offensive cyber forces, and the resources that technology companies devote to protecting their platforms from becoming digital havens for jihadis, it would seem easy to predict quick, satisfying victories in the cyber battle against bands of ill-funded terrorists. The reverse turns out to be true. "It's the hardest fight we face," one senior military official told me. Blow up a safe house in Pakistan or a missile base in Syria, and the result is rubble. Aim at the servers sending out beheading videos or recruiting messages, and the videos and messages just reappear elsewhere a few days later.

"It's almost never as cool as getting into a system and thinking you'll see things disappear for good," said Joshua Geltzer, the senior director for counterterrorism at the National Security Council under Obama. By the time Obama left office, the question of whether Cyber Command had pursued ISIS with enough vigor became so fraught that there was a movement inside the Obama administration to fire Adm. Rogers.

But while Washington was struggling to understand how to go on the offense against groups that were using social media as a way to organize attacks, Silicon Valley was still unable or unwilling to face the extent of the problem. For years the world's most brilliant technologists convinced themselves that once they connected the world, a truer, global democracy would emerge. They rejoiced when Twitter and WhatsApp made the Arab Spring possible, and were convinced they had built the weapon that would tear down autocrats and beget new, more transparent democracies.

But over time a harsher truth has emerged. Those same networks became ISIS's most potent tool. They were exploited by Russian trolls and the political targeteers at Cambridge Analytica to manipulate voters. And the subsequent call for a new kind of cyberspace—where we understand the real identities of everyone we are dealing with on the web—delighted the Chinese and the Russians. What better way to hunt down dissidents and doubters, and break up the political opposition?

Meanwhile, the tech companies became gradually aware of another international threat to their future: China's carefully laid-out plan to become the world's dominant economic and technological power by

2049, the hundredth anniversary of Mao's revolution. To accomplish this goal, Beijing developed a new strategy—one in which Chinese investors, rather than the venture capitalists of Sand Hill Road, were quietly becoming a critical source of cash for a range of new start-ups.

Suddenly, the valley's billionaires discovered they needed something that they had never really thought about before: a foreign policy.

IN THE SPRING of 2016, under pressure to reverse the Islamic State's expansion in Syria and Iraq, the Pentagon announced for the first time that it was declaring cyberwar on a foreign entity.

"We are dropping cyber bombs," Robert Work, the usually staid deputy secretary of defense, told reporters in a bit of hyperbole that raised the eyebrows of his Pentagon colleagues. "We have never done that before."

They had, of course, but without announcing it publicly. The job had been handed to Gen. Paul Nakasone, who after Nitro Zeus had moved on to run the army's cyber operations and was already rumored to be in line as the next head of the NSA and Cyber Command. Soon, "Joint Task Force Ares"—a combination of forces designed to go after ISIS networks—was created in Florida at Central Command, which directs American military operations in the Middle East. Cyber Command mission teams poured into MacDill Air Force Base and other Central Command posts, joining up with the more traditional military units that were operating against ISIS.

The goal of the new campaign, I was told in a series of briefings, was to disrupt the Islamic State's ability to spread its message, attract new adherents, circulate orders from commanders, and carry out day-to-day functions, including paying its fighters. Even Obama entered the fray, emerging from a lengthy meeting at the CIA about operations against the Islamic State and declaring, as part of a description of the strategy, that "our cyber operations are disrupting their command-and-control and communications." Clearly the administration went public for only one reason: to rattle ISIS commanders and increase their paranoia that someone was inside their communications and perhaps manipulating their data. The theory was that potential recruits would be deterred if they began to worry about the security of their communications.

But while these pronouncements sounded impressive, the results were not. Obama's top aides were growing increasingly impatient about how slowly Cyber Command was finding all the dark corners of the Internet where ISIS was hiding its digital caches of recruiting and training material, and how quickly it all resurfaced when knocked out. "The Internet is a big place," James Clapper told me, "and ISIS is very astute and sophisticated about how they used it," often stashing their materials in the cloud via servers located in Germany and elsewhere.

No one was more impatient with the pace of progress than Ashton B. Carter, the tech-savvy defense secretary, who was pouring his all into an ISIS strategy. Carter was a physicist and had been a major force in pushing the Pentagon to develop far greater cyber capabilities and the doctrine to match. But he had diminishing patience for Rogers, who he thought was not putting enough resources or creativity into the problem of knocking the terror group offline—and keeping them offline.

"Ash was holding meetings every few weeks, even traveling out to Fort Meade, pounding on them to do more," one senior official involved in the tense standoff told me. By the summer of 2016, the tension between Carter and Rogers had grown so intense that the defense secretary was looking to replace him—chiefly because of the continued leaks of cyberweapons from inside the NSA's Tailored Access Operations Unit, but also because of the lack of progress in the digital war against ISIS. Clapper concurred, Pentagon and intelligence officials say, but lacked Carter's enthusiasm.

"It was debated," a top White House official told me later about the movement to replace Rogers. "But we concluded time was so short we probably couldn't even get his replacement through."

Carter kept pressing to take ISIS offline, and after many delays the last big cyber operation of the Obama years began three months behind schedule—in November 2016, just as questions about Russia and its election influence were dominating the post-election headlines. "Operation Glowing Symphony," as it was code-named, would be the largest cyber effort against ISIS and one of the last big cyber operations that Obama approved in the Situation Room.

The idea was to combine the best skills of the NSA and US Cyber Command, steal the passwords of several Islamic State administrator accounts, and then use them to trigger chaos in the networks—blocking out some fighters, deleting some content, altering data to send convoys to the wrong place. It didn't sound especially high-tech. And at first it looked successful because some battlefield videos disappeared. Clearly ISIS fighters were distracted, and disturbed.

But the effects were fleeting; the videos started reappearing elsewhere. And the ISIS commanders had backup systems, and quickly switched networks, using servers spread out over three dozen or so countries. One senior official recalled that Cyber Command would show up "with PowerPoints about all the setbacks they had caused, but they couldn't answer the simple question: 'How much lasting effect did you have?'"

That prompted the outbreak of another debate: Could Cyber Command go after the servers in three dozen nations—including in Germany—without telling allies that they were about to conduct offensive cyber operations against ISIS inside their national networks? "They wanted to use another country's infrastructure, and the question was whether we had to tell the countries first—and if we did, whether the whole thing might leak out," the official said. The debate dragged on for weeks; Obama ultimately decided that the intelligence agencies should seek permission from allied countries because the United States would be outraged if we discovered the British or the French or the South Koreans were using American networks to conduct military activity.

Ultimately, time ran out and the program was handed over to the

Trump team. In 2017 ISIS was largely driven from Syria and Iraq, and Carter and his team deserved much of the credit; it was their plan. But once he left the Pentagon, Carter wrote a blistering assessment of how the cyber operations had played out.

"I was largely disappointed in Cyber Command's effectiveness against ISIS," he wrote in late 2017, in a remarkably candid account. "It never really produced any effective cyberweapons or techniques. When Cybercom did produce something useful, the intelligence community tended to delay or try to prevent its use, claiming cyber operations would hinder intelligence collection. This would be understandable if we had been getting a steady stream of actionable intel, but we weren't."

"In short," he added, "none of our agencies showed very well in the cyber fight."

Carter's critique was about more than just how Cyber Command did against ISIS: many in the Pentagon, and certainly at the NSA, questioned the overall performance of America's newest fighting unit. "There is just not much capacity there," one senior Pentagon official said to me in 2017, trying to answer the question of why the attacks on ISIS had gone so slowly.

In fact, eight years after its creation, Cyber Command was still overly dependent on the technology and tools of the NSA; as one member of a key Cyber Mission Team told me, "Most of the time, we were just using their stuff." Partly that was to be expected of a start-up venture, but partly it was because the rest of the military didn't know exactly how to arm and train soldiers who worked at keyboards all day.

That was evident in how Cyber Command was staffed. Hundreds, then thousands, of enlisted women and men and their officers rotated through the Cyber Mission Forces for two-year stints, learning how to defend Pentagon assets in cyberspace, or how to provide specialty support to Pacific Command or Central Command as they took on the Chinese or the Iranians. But it turned out that two years was barely enough time to learn the intricacies of breaking into foreign computer networks and executing operations. That process could take years of patient work, and frequently the members of the 133 Cyber Mission Forces had moved on before their operations came to fruition. Worse yet, when they returned to the air force or the army or the marines, they were frequently assigned to jobs in which their newly acquired cyber skills played little or no role.

In contrast, the civilians working next door at the NSA spent years developing tools, learning the insides of Russian or North Korean or Iranian networks, and implanting their malware. Often they treated these "implants" like prized bonsais, to be watered, nurtured, and cared for. The culture of the NSA was far more risk-averse, and to them, Cyber Command offensive units were mostly interested in blowing things up, which exposed and rendered useless the implants the NSA had so carefully hidden.

Moreover, the early hopes that Cyber Command would prove to be the military's new Special Operations Forces turned out to be more hype than reality. "They simply didn't run at the tempo of Special Forces—they weren't hitting foreign networks every night the way the Special Forces hit houses in Afghanistan," said one senior official who was dealing with both the NSA and Cyber Command. "And so they didn't have a lot of opportunity to learn from their mistakes."

The fact of the matter was that the US was simply not conducting major cyber operations against foreign adversaries at anywhere near that pace; at most, the Cyber Mission Forces assigned to conduct offensive operations were doing just a few each year, every one requiring presidential authorization. The result was that Cyber Command came to resemble its parent, Strategic Command, which watches over America's nuclear forces: they spent a lot of time training, debating doctrine, establishing procedures for operations, and playing out scenarios.

None of those scenarios, it turned out, involved what would happen if a foreign power tried to manipulate an American election.

ALEX STAMOS, the blunt, bearded, whip-smart security chief for Facebook, had a simple explanation for why the world's biggest purveyor of news and communications didn't see the propaganda that the Internet Research Agency and other Russian groups were distributing to influence the 2016 election: they simply weren't looking.

"The truth is that it was no one's job to look for straight-up propaganda," Stamos told me in February 2018, as the world began to fall in on the company that once had viewed itself—with one part hubris, one part blindness—as a force for the spread of democracy and the bane of dictators.

"We were looking for traditional abuse of the platform," he said at the Munich Security Conference, an annual gathering of foreign ministers, national-security officials, and think-tankers that, in 2018, was consumed by the new weaponry of cyber and social media. "We missed it because we weren't looking."

Stamos knew the vulnerabilities of complex systems cold—and had little time or patience for executives who didn't like to hear stark assessments of why their ideas wouldn't work. Among those who discovered how quickly Stamos could get in someone's face was Adm. Michael Rogers, in his early days at the National Security Agency. In February 2015, I was at a cybersecurity conference where Rogers was speaking, in his usual measured terms, about balancing the need for encrypted communications with the government's need to be able to decrypt conversations among terrorists, spies, and criminals.

Stamos, then the chief security officer at Yahoo!, grabbed a microphone and repeated to Rogers the argument that creating a back door in a communications system was akin to "drilling a hole in a windshield." The entire structure would be so weakened, he suggested, that it would destroy the concept of secure communications. When Rogers went into his routine about balancing interests again, Stamos kept pushing. The video of the event quickly went viral.

Over time, though, Stamos's insistent voice grated on the top leadership at Yahoo!, especially when he pressed for such full encryption of data that Yahoo! itself would not be able to decrypt the communications on its own platform—mimicking what Apple had done with the iPhone. Of course, if Yahoo! could not pluck out keywords from its customers' searches and communications, it couldn't make money by selling advertising and services that catered to them. There was no way that Marissa Mayer, the struggling company's chief executive, was going to choose that much privacy protection over the revenues arising from harvesting the habits of Yahoo! users. Stamos soon left to become Facebook's chief security officer—where he would also run afoul of the leadership.

The Facebook that Stamos joined in 2015—eleven years after the company's storied founding—still thought of itself as a huge pipeline for the vast transmission of content, but not as a publisher. Its business plan was based on the assumption that it would not exercise editorial judgments; instead, like the telephone company, it would carry content but not edit it. Naturally, this was a false analogy: From the start, Facebook made its money not by selling connectivity, but by acting as the world's seemingly friendly surveillance machine, then selling what it learned about users, individually and collectively. The old phone companies never did that. As my colleague Kevin Roose wrote, "Facebook can't stop monetizing our personal data for the same reason Starbucks can't stop selling coffee—it's the heart of the enterprise."

Yet the idea that Facebook and its competitors could pursue that strategy and ignore the content of what was appearing on their platforms—and thus avoid editing on a massive scale—lay somewhere between naïveté and delusion. Phone companies had to crack down on telephone fraud; network television couldn't broadcast pornographic films. Even Netflix confronted limits. So over time Facebook was forced to keep revising its "terms of service," defining exploitative, racist, or illegal activity (selling drugs, gambling, or prostitution, for example) that would not be permitted. But the company never referred to these as editorial decisions; instead, they were "community standards" that would force it to disable accounts or alert law-enforcement authorities if there was a "genuine risk of physical harm or direct threats to public safety."

It all had a feel-good sensibility to it—until Facebook tried to define what those policies meant in real life. It began with simple things. Soon it got very complicated.

Child pornography was easy; it was banned early in Facebook's history. Then parents of newborns discovered that the site was taking down pictures they were sharing of their own babies in the bath. If they reposted them, their accounts were disabled. Then, in 2016, came the first moment when Facebook executives discovered they had to make a news judgment, because, as it turns out, algorithms have no sense of history. Norway's major daily, *Aftenposten,* put on a photography exhibit and posted on Facebook the iconic Vietnam War photograph of a young girl running down the road, naked, to escape napalm and mayhem. The photo won the 1973 Pulitzer Prize. Naturally, the picture was immediately banned by Facebook's algorithms. *Aftenposten* called the company out, and within a day a clearly ridiculous deletion was reversed, after a series of hurried conference calls among Facebook executives. It was not a hard decision.

But that conference call was, essentially, the first time senior Facebook managers had to think like news editors, balancing their rules against history, artistic sensibility, and, most important, news judgment. It was the moment when they realized that no algorithm could do the job. When I said that to a senior official at the company, he grimaced and asked, "You think there will be more?"

His question reflected how deeply the company was in denial about what was coming. Facebook executives, along with their colleagues at Google, celebrated when their creations helped organize students in Tahrir Square to oust President Hosni Mubarak, and Libyans to overthrow Muammar Gaddafi in Libya. "The Arab Spring was great," Alex Stamos said to me. "Glory days." But time was not on the side of the democrats. Little thought had been given to what would happen when the world's autocrats and terrorists caught on, and the degree to which the same platforms enable social control, brutality, and repression.

It started with the beheading videos in the early 2000s. And the captured pilot burned alive in a cage. On Twitter, ISIS operatives created the Dawn of Glad Tidings app for their followers to download, which allowed them to send out mass messages detailing recent attacks, complete with coordinated images and hashtags. Social-media companies found that new accounts popped up faster than old ones could be found and killed off.

Of course, the decision to take down the beheading videos was an easy one; they violated the "terms of service." Yet the companies quickly ran into the same problem the NSA did: There are lots of places to hide on the Internet. ISIS had placed perfect digital copies of their library of horror and recruitment videos all around the world. A predictable arms race ensued. YouTube automated its review systems to speed the process of bringing down videos.

It became an unwinnable game of digital whack-a-mole. As Lisa Monaco, Obama's homeland security adviser, said, "We are not going to kill our way out of this conflict. And we are not going to delete our way out of it either."

In 2017, Facebook and others rolled out an impressive technological fix, a "digital fingerprint" for every image, created by turning every beheading photograph or exploitative picture of a child into a black-and-white image and then assigning each pixel a numeric value based on its contrast.

"Let's say that somebody uploads an ISIS propaganda video," explained Monika Bickert, a former prosecutor who became Facebook's representative to governments around the world. "With the fingerprint, if somebody else tries to upload that video in the future we would recognize it even before the video hits the site." But discerning motive, she concedes, requires human review. "If it's terrorism propaganda, we're going to remove it," she told me. "If somebody is sharing it for news value or to condemn violence, that's different."

Google, meanwhile, tried a different approach: "Google Redirect," an effort to send people who searched for ISIS propaganda or white-supremacist content to alternative sites that might make them think twice. To make it work, Jigsaw, the company's New York–based think-tank and experimental unit, interviewed scores of people who had been radicalized, trying to understand their personality traits, starting with their distrust of mainstream media.

"The key was getting them at the right moment," Yasmin Green, who helped spearhead the movement, told me. It turned out there was a small window of time between when potential recruits develop an interest in joining an extremist group and when they make the decision. Rapid exposure to the personal testimony of former recruits who escaped from the brutalities of life inside ISIS, and describe it in gory detail, was far more likely to dissuade new recruits than lectures about the benefits of a liberal view of the world. So were the accounts of religious leaders who could undercut the group's argument that it was following the Koran.

"Our job is to get more and better information in the hands of vulnerable people," Green told me.

But that job requires the world's pipeline providers to create, delete, and choose. In short, they had become editors. Just not fast enough to stay ahead of the Russians.

MARK ZUCKERBERG QUICKLY came to regret his dismissive words, six days after Donald Trump's election, that Facebook had nothing to do with it.

"Personally I think the idea that fake news on Facebook, which is a very small amount of the content, influenced the election in any way—I think is a pretty crazy idea," Zuckerberg wrote. "Voters make decisions based on their lived experience. I do think there is a certain profound lack of empathy in asserting that the only reason someone could have voted the way they did is they saw some fake news. If you believe that, then I don't think you have internalized the message the Trump supporters are trying to send in this election."

Nine days later Zuckerberg was in Peru, at a summit that President Obama was also attending. The president took him into a private room and made a direct appeal: He had to take the threat of disinformation more seriously, or it would come to haunt the company, and the country, in the next election. Zuckerberg pushed back, Obama's aides later told me. Fake news was a problem, but there was no easy fix, and Facebook wasn't in the business of checking every fact that got posted in the global town square. Both men left the meeting dissatisfied.

By the time Zuckerberg spoke, however, Alex Stamos and his security team were nearing the end of an excavation of Facebook history, digging into the reports of how the Russians had used ads and posts to manipulate voters. As Stamos dug, he began to run into some quiet resistance inside the company about going further. His study was delivered to the company's leadership on December 9, 2016, laying out the Russian activity his group had found. But when it was ultimately published four months later under the bland headline "Information Operations and Facebook," it had been edited down to bare essentials and stripped of its specifics. Russia was not even mentioned. Instead the study referred to anonymous "malicious actors" that were never named. It played down the effects, confusing volume with impact. "The reach of known operations during the US election of 2016 was statistically very small compared to overall engagement on political issues."

Then it concluded: "Facebook is not in a position to make definitive attribution to the actors sponsoring this activity."

In fact, they had a pretty good idea by April that "Fancy Bear," the Russian group directed by the GRU, was behind some of the Facebook activity. The company turned much of the evidence about ads over to Senate investigators in September. But it was wasn't until the Senate published some of the juiciest examples—from ads designed to look like they were part of the Black Lives Matter movement to another in which Satan is shown arm-wrestling Jesus and saying, "If I Win Clinton Wins!"—that the company was forced to admit how much propaganda ran on its site.

"The question was, how had we missed all this?" one Facebook executive told me. The answer was complex. The ads amounted to very little money—a few hundred thousand dollars—and it was not obvious they were coming from Russia. Later, after the ads were discovered, Facebook's lawyers began to worry that if the ads and posts of private users were made public—even those created by Russian trolls—it could violate Facebook's own privacy policies.

In September 2017, ten months after the election, the company finally began to concede the obvious. It said those who had manipulated Facebook "likely operated out of Russia," and it turned over 3,000 of these ads to Congress. It had found evidence that the Internet Research Agency created 80,000 posts on Facebook that 126 million people may have seen—though whether they absorbed the messages is another question. But Facebook was still insisting it had no obligation to notify its users that they had been exposed to the material.

"I must say, I don't think you get it," Sen. Dianne Feinstein, the California senator who was usually a great advocate of her state's economic champions, said during the resulting hearing. "What we're talking about is a cataclysmic change. What we're talking about is the beginning of cyberwarfare." That wasn't exactly right: Whether it was warfare depended on how you define the term. And if it was cyberwar, it wasn't the beginning, by a long shot.

By the spring of 2018, Facebook was reeling. Additional disclosures that it had given access to its user profiles to a scholar in 2014, who in turn massaged the data and used it to help Cambridge Analytica, a London company that targeted political ads for the Trump campaign, forced Zuckerberg to a new level of contrition. The problem was that Facebook's users had never signed up for having their lives and predilections examined, then sold, for such purposes. "We have a responsibility to protect your information," Zuckerberg declared in ads and a series of carefully scripted television interviews. "If we can't, we don't deserve it."

The more telling concession came out of France, where the company began announcing a radical experiment. It would begin to fact-check photos and videos around elections, it said—just as news organizations have done for decades. Sheryl Sandberg, the company's chief operating officer and one of the few executives who had serious Washington experience, offered the most candid assessment: "We really believed in social experiences," she said. "We really believed in protecting privacy. But we were way too idealistic. We did not think enough about the abuse cases." Yes, it was naïveté. But it was a naïveté that helped drive immense profits—and blinded the company's top executives to the consequences of how the information it entrusted to it by Facebook's users could be abused.

BEFORE HE BECAME the Pentagon's resident technology scout and venture capitalist in Silicon Valley, Raj Shah spent twelve years flying an F-16 around Afghanistan and Iraq. Much of that time he wondered why a $30 million aircraft had worse navigation systems than a Volkswagen.

The mapping technology was so ancient that it did not, at a glance, show pilots how close they were to national borders, or the features of cities and towns below them. Worse yet, Shah told me one day as we walked around his Pentagon-created start-up—called DIUx for "Defense Innovation Unit, Experimental"—"I had no way of knowing if I was flying into Iran" by mistake, a potentially fatal error.

When Shah was back home on leave, he often rented a Cessna and zoomed around with a $350 iPad mini strapped to his thigh. With an app called ForeFlight on the mini, he could see exactly where he was, every feature of the landscape below. He could look at the map or a satellite photograph. It tracked him with near precision. "I knew exactly where I was." As he thought about how he had better mapping on a beat-up iPad than in America's workhorse fighter, he could come to only one conclusion: "This is screwed up."

To Shah, the experience exemplified what was wrong with how the Pentagon equipped war fighters. Systems designed with 1970s technology couldn't be easily upgraded, because the process of testing to make sure they are "military grade" takes years—by which time the technology is out of date. "This is why we have fifty-year-old aircraft carriers," Shah said, "with thirty-year-old software."

When Shah's flying days were coming to an end, he landed back in Silicon Valley, "where everything runs on speed." From his new perch, the Pentagon's mode of operating seemed even more ridiculous. "Who wants a four-year-old phone?" he asked me one day.

Shah had some allies in holding this view, including Ashton Carter, who had spent months in Silicon Valley between the time he was deputy secretary of defense and when Obama brought him back to serve as secretary in February 2015. Carter and his chief of staff, Eric Rosenbach, another major architect of the Pentagon's cyber efforts, were determined to use their two years to change the culture of the Pentagon. They sponsored a "Hack the Pentagon" challenge, setting up prizes for hackers who could find holes in the security of Pentagon programs (the star of the competition, naturally, was an eighteen-year-old high school senior whose mother dropped him off at the Pentagon to pick up his reward.) They tried to get Silicon Valley technologists to spend a year or so in government service, with only partial success. One year was, as one of the experimenters put it, "just long enough for a massive clash of cultures, but not long enough to get much done."

But the most important experiment was Carter and Rosenbach's creation of DIUx. Ultimately it placed about fifty officers and civilians in the heart of Silicon Valley with the explicit mission of finding existing cutting-edge technologies that could be rapidly given to war fighters, Special Forces operators, marines, or army personnel.

Shah and Christopher Kirchhoff, a veteran of the Obama White House and the office of the secretary of defense, were brought in as lead "partners"—the word indicating that the Pentagon was intent on blending in with the Silicon Valley natives. They were a good mix, but Shah knew from his own experience how hard it would be to get companies in Silicon Valley to participate in any effort sponsored by the Pentagon. First, there was the post-Snowden hangover. But there was also a very practical fear of getting wrapped up in the deadening, regulation-heavy Pentagon bureaucracy. "When I was getting a business start-up going, my investors wouldn't let me even talk to the military," he said. "They complained it took years to close a deal, and even longer to get paid. You'd go bankrupt just waiting for them to make a decision."

As with any new venture, Shah needed cool office space. But while the Pentagon was ready to be a player in Silicon Valley, it wasn't ready to pay the sky-high rents. So DIUx commandeered space in an old military building in Mountain View, just on the edge of the Google campus. Autonomous Google cars buzzed past the headquarters constantly.

But what DIUx really needed were some quick wins—off-the-shelf products that could save lives and demonstrate to the Pentagon that its old, slow acquisition methods left troops vulnerable and made war fighting harder than it needed to be. "The key is using technologies that are already available," Shah told me, "and making the modifications we need for a specific military purpose."

Soon his offices were filled with possibilities. Among the most promising first finds was a quad-copter—a tiny, autonomous helicopter that had been developed to help construction companies. The devices fly indoors and up stairwells to confirm, with laser measurements, that every wall has been built to precise specs. Upon seeing the copter, Shah and his team suspected it could prove immediately useful in Iraq and Afghanistan. When Special Forces were getting ready to clear out a house full of suspected militants, they needed to know where everyone was—and whether the place had been booby-trapped with explosives. Though it took longer than it should have, Shah got prototypes into the hands of troops on the ground.

But the most urgent project—and the one that ran afoul of the Pentagon and its biggest and most powerful contractors—involved launching a new kind of tiny satellite over North Korea to keep track of the missiles Kim Jong-un was placing on hard-to-track mobile launchers. The giant spy satellites the United States has relied on for decades were a prime example of how the Pentagon was dependent on technology that was outdated the day it was launched: they cost billions, took years to design and build, and were kept in orbit for years, until they fell out of the sky.

Shah and Kirchhoff envisioned something completely different: tiny, backpack-sized, inexpensive civilian satellites that had been developed to count cars in Target parking lots and monitor the growth of crops. They were launched in clusters, and would stay in orbit just a year or two. But they were also so cheap that when they fell out of the sky, the Pentagon could simply launch the newer, higher-resolution replacements. This would likely provide the kind of coverage needed to execute a new military contingency plan called "Kill Chain," in which satellites would identify North Korean launches, or nuclear facilities, and provide the targeting data to destroy them preemptively if a conflict seemed imminent.

The urgency arose from the fact that the American coverage of North Korea from space was (and remains) terrible—the United States had eyes on the country less than 30 percent of the time. (The exact figure is classified.) William Perry, the former defense secretary, told me that if the North Koreans rolled out one of their new missiles, "there's a good chance we'd never see it."

"Kim Jong-un is racing—literally racing—to deploy a missile capability," Robert Cardillo, the director of the National Geospatial-Intelligence Agency, told me in mid-2017. "His acceleration has caused us to accelerate."

But the Pentagon was still not moving fast enough. The North was adding missiles and missile launchers, which could be hidden in caves and rolled out minutes before launch, faster than the defense bureaucracy could stomach the idea of letting small firms like Capella Space Corporation, a start-up named after a bright star, threaten the traditional contractors. Capella was one of a number of small satellite firms that had slashed the cost of space-based radars that can see through clouds, rain, snow, camouflage, and foliage, and pick out changes in the ground elevation that can point to hidden tunnels. Just ten years ago, building a constellation of those satellites was estimated to cost more than $94 billion. Now, Shah believed, it could be done for tens or hundreds of millions. It seemed like a no-brainer, especially with the warnings that the North was investing in solid-fuel missiles that could be rolled out of caves quickly, and launched in hours or minutes. Early detection was key.

Yet as Shah traveled back and forth between Silicon Valley and Washington, trying to get the satellites launched on an accelerated scale, he ran into one impediment after another. Members of Congress didn't like the way DIUx was using a loophole in acquisition rules to short-circuit the usual, tortuous process of getting R&D projects approved. The big satellite makers, while faking enthusiasm, felt their multibillion-dollar contracts might be threatened by start-ups they had barely heard of. They were a powerful lobby.

"We're trying to do something very different," Shah told me with considerable understatement. "And that always ruffles people."

The bureaucratic reluctance in Washington to move at entrepreneurial speeds was bad enough. But as Shah and Kirchhoff moved around the Valley, exploring possible new technologies, they kept running into a competitor with the same vision, less reluctance, and a much bigger budget: it turned out the Chinese had an informal kind of DIUx of their own under way—on American soil.

"I don't think anyone realized the extent of it," Kirchhoff told me one evening. "In the old days the Chinese bought up companies," until the United States began to reject many of those deals on national-security grounds. So the Chinese were taking another path to the same destination. They opened up their own venture-capital funds in the Valley, as well as investment funds that could take a minority position in new companies—not big enough to trigger a review, but enough to get an early understanding of the technology.

But no one could quantify China's involvement—there were no concrete numbers that gave a holistic sense of the Chinese investment strategy. So Shah and Kirchhoff found a savvy executive who under-stood Silicon Valley—Michael Brown, who had run Symantec, the company that identified so many of the details of Stuxnet—and asked him to write an unclassified report on it. They hoped this would wake Washington up to the new ways that the Chinese were scooping up Silicon Valley's most critical technologies for both their state-owned companies and their own military. Brown quickly joined forces with Pavneet Singh, a former National Security Council and National Eco-nomic Council staffer who had prepared Obama for his meetings with President Xi.

"The first thing I learned is that the Chinese have done a beautiful job mirroring Silicon Valley," Brown told me. Baidu is the answer to Google. Alibaba is the Chinese Amazon. Tencent is known for apps used by two-thirds of all Chinese for messaging and money transfer, apps like WeChat and QQ. By one venture capitalist's measure, the Chinese spend 1.7 billion hours a day on Tencent apps.

But the second thing Brown learned was that the Chinese were es-sentially doing what DIUx was doing—investing in "early-stage" com-panies. They were just doing it on a far larger scale than the Pentagon had in mind.

What made the strategy so brilliant was that the Chinese were flying in under the radar. When they bought an entire company, it triggered an official review in Washington, from a little-known, little-understood group called the Committee on Foreign Investment in the United States. It could recommend that the president block any sale on national-security grounds. And both Obama and Trump did so, several times.

But there were no rules about taking a minority stake in a com-pany, which would give the investors an early, privileged look at the technology. This alarmed Brown, who knew that the Chinese investors were getting a detailed look even at companies and technologies that they turned down for investment. Nobody had ever imagined this pos-sibility when the foreign-investment rules were written decades ago:

venture capital barely existed, and it was unimaginable that the Chinese would be players in the game.

"People keep saying to me, 'You are overreacting—remember Japan,'" said Brown, referring to the time in the late 1980s and early '90s when some feared that the United States would become a "techno-colony" of the Japanese. But Brown became convinced this was different. "Japan was a staunch ally. It was never a military rival in that time. It never had a chance of challenging the US economically. And we had shared values." In China's case, none of that applies. (It is a sign of the change in economic fortunes that Japan, the country the United States worried most about in the '80s and '90s, provided only $13 billion of the venture capital in the DIUx charts—only a little more than half of what the Chinese were spending.)

The DIUx report's findings, which began circulating confidentially around Washington in the spring of 2017, were astounding. They demonstrated that even while the Chinese were paring back on *stealing* the fruits of American industry—Obama's agreement with Xi had begun to have some effect—they had found many perfectly legal ways to *invest* in it. A government that still gave lip service to communism had figured out venture capitalism—and concluded it was the shortest path to get the technologies the country needed.

The numbers that Brown and Singh gathered, all from public sources, told the story. China participated in more than 10 percent of all venture deals in 2015, the report found, focusing on early-stage innovations critical to both commercial and military uses: artificial intelligence, robotics, autonomous vehicles, virtual reality, financial technology, and gene-editing. When they broke down who was investing in US-based venture-backed companies between 2015 and 2017, American investors ranked first, with $59 billion in investment. Europe was second, with $36 billion. And China was right behind, with $24 billion.

Some of the biggest direct investments came from Baidu and Tencent, but there were also a surprising number from venture-capital firms with Western-sounding names—West Summit Capital and Westlake

Ventures—that were wholly Chinese-owned. "They are private actors," Brown told me, "but always acting with the approval of the Chinese government."

Xi Jinping made it clear he was completely behind the strategy by sending subtle messages. In the fall of 2017, at the 19th Party Congress, Xi talked about how he was targeting these strategic areas. In his New Year's speech in 2018, Xi had books on artificial intelligence placed strategically behind him, so they would be caught on camera. Chinese investors in the Valley didn't need to see the speech to get the message: they did eighty-one deals in American artificial-intelligence companies between 2010 and 2017, worth $1.3 billion. More than a third of that—over $500 million—was spent in 2017 alone.

Brown and Singh's DIUx report was soon in the hands of Gen. Paul Selva, who held the vice chairman post at the Joint Chiefs of Staff, the job once occupied by James Cartwright. General Selva had encouraged the study and used it to sound the alarm inside the Pentagon. But the report arrived in the early days of the Trump presidency, and rather than serve as a call for the United States to think in Chinese terms about how best to invest in research and development—and how to integrate those investments with defense projects—the report became another excuse for Trump's calls for protectionism. The Trump economic team somehow convinced itself that it could cut the Chinese off entirely from investing in sensitive new technologies in the United States—even while China held $1.2 trillion in US debt.

By early 2018, Trump was looking for new ways to block Chinese investment, and even non-Chinese investment that might help Beijing. He stopped a Singapore company, Broadcom, from buying Qualcomm, a maker of vital but specialty chips that are used by US Special Forces, among other military units. The stated fear was that Broadcom, while not itself Chinese, would not invest heavily in research—and Huawei and other Chinese firms would benefit.

Anthony Balloon, an international lawyer concentrating on China, told my *Times* colleagues that this was a turning point in a growing technology war: "There is now a recognition in government that foreign investors, particularly from China, are getting more and more sophisticated on how they get access to technology in the US." The message was clear: the United States would look anew even at minority-stake investments and other forms of Chinese capital.

And American officials became explicit about banning Chinese technology that they believed could give Beijing a back door into US networks. In March 2018, when Paul Nakasone was finally nominated to run the NSA and US Cyber Command, he told Congress with a smile that he would never even use a Huawei phone; around the same time, Best Buy stopped selling them. What Nakasone didn't say was that even though Huawei had emerged as the world's largest builder of networking equipment—and was wiring most of Asia and some of Europe—the NSA had quietly banned AT&T and Verizon from even allowing Huawei to bid on building parts of America's 5G network. The companies, and even some in the intelligence agencies, argued that this was shortsighted: if Huawei bid, they pointed out, it would have to allow the United States to pick apart the details of how its networks are built. American officials shrugged and said no.

It was a remarkable progression. What had begun early in the Obama administration as a fear that China was using cyber techniques to steal American technology for the benefit of its state-owned companies had turned into a much larger technological cold war. The Chinese weren't stealing as much. Instead, they were buying into America, perfectly legally. And the United States was struggling to figure out how to stop them without rejecting the principles of a free, global market.

THE REAL WARNING contained in the DIUx document was not about what the Chinese were doing in Silicon Valley, but about what they were working on at home.

The Obama-era battles over the theft of intellectual property captured the headlines, and in the spring of 2018 President Trump revived the issue as part of his justification for the threat of massive tariffs against Chinese goods. While the pace of theft had lessened, without question many of China's large and small firms were still looking to acquire research, development, and product blueprints by any means—legal or illegal. But the equally big concern was China's focus on becoming the world's leader in artificial intelligence by 2030, and its huge investments in the technology it needs to accomplish that task. A single, massive campus in Hefei, in Anhui Province, captures the scope of that effort: there, China is constructing a $10 billion research hub called the National Laboratory for Quantum Information Sciences, the centerpiece of its quantum computing effort.

Quantum computing—the ability to take calculations to warp speed by using photons instead of manipulating ones and zeros the old-fashioned way—holds the key to cracking any encryption by brute force. If successful, it could result in developing secure communications links, and navigation systems that do not rely on global positioning satellites, which can be jammed by an adversary or used to locate a hidden submarine. The Chinese have already tested quantum satellites.

"The question is, how should the United States respond to this challenge?" Robert Work, the former deputy secretary of state who pushed the Pentagon into competing in this arena, told my colleague Cade Metz. "This is a Sputnik moment."

Work's analogy captured a critical truth: any breakthroughs produced by the concentrated Chinese effort will flow directly to the country's military might. An equivalent to the Silicon Valley/Washington divide, which bubbled along before Edward Snowden's disclosures and re-erupted in the battles over the government's effort to get a back door into encrypted systems, does not exist in China.

Yet in the United States, the divide is widening. The Cold War model, in which breakthroughs in American military technology and the space program flowed to the commercial sector, is gone forever. The reverse model—using the skills of Silicon Valley to create the next-generation weapons—has run headlong into political and cultural opposition.

"Even if the US does have the best AI companies, it is not clear they are going to be involved in national security in a substantive way," said Gregory Allen of the Center for a New American Security. The effects are already visible: the military edge the United States has grown accustomed to since World War II is eroding.

The post-Snowden opposition to cooperating with the military broke out anew in the early spring of 2018 on the Google campus, just blocks from the DIUx headquarters. News of Google's plans to participate in "Project Maven," a pilot Pentagon program that uses artificial intelligence techniques to process "wide area motion imagery" that detects moving vehicles and moving weapons systems, sparked an internal uprising. As word spread through the company, thousands of Google's employees signed a letter that opened with this declaration: "We believe that Google should not be in the business of war." The letter rejected the company's assurances that its work was not helping the Pentagon "operate or fly drones" or launch weapons. Google employees rightly saw those assurances as a dodge—the project might not be aiding current weapons, but clearly the Pentagon intended for the results to be incorporated into future weapons.

"The technology is being built for the military," the employees wrote, "and once it's delivered it could easily be used to assist in these tasks." Then came the punch line: "By entering into this contract, Google will join the ranks of companies like Palantir, Raytheon and General Dynamics." The statement concluded by urging the company to draft a "clear policy stating that neither Google nor its contractors will ever build warfare technology."

The rebellion was not limited to Google. At the same moment, Microsoft was quietly getting dozens of other firms to sign on to an agreement that they would never knowingly help any government—the United States or its adversaries—build cyberweapons for use against "innocent civilians." They vowed to help any country that finds itself attacked.

At the core of these uprisings is a concept of corporate identity that is the complete reverse of the Cold War. Raytheon and General Dynamics flourished because they were part of an American defense establishment that armed the Western alliance. They were serving governments, not consumers, and so of course they willingly picked a side.

Google and Microsoft do not share this view. Their customers are global, and the bulk of their revenue comes from outside the United States. They view themselves, understandably, as essentially neutral— loyal to the customer base first and individual governments second.

Washington, in contrast, still views them as "American companies," beneficiaries of American freedoms. In the Pentagon's view, their expertise and technology should flow first to defend the nation that allowed them to form and flourish. These are two completely distinct worldviews, which, at least in peacetime, will never be aligned.

CHAPTER XII

LEFT OF LAUNCH

MARY LOUISE KELLY (NPR): Is there a Stuxnet for North Korea?
JOHN BRENNAN (CIA DIRECTOR UNDER PRESIDENT OBAMA):
[Laughter] Next question.

—*December 2016*

I N THE SPRING of 2016, North Korea's missiles started falling out of the sky—if they even made it that high.

In test after test, Kim Jong-un's Musudan missile—the pride of his fleet—was exploding on the launch pad, crashing seconds after launch, or traveling a hundred miles or so before plunging prematurely into the Sea of Japan. For a missile that Kim imagined would enable him to threaten the American air base on Guam and form the technological basis for a larger missile that could reach Hawaii or Los Angeles, the failures were a disaster.

All told, Kim Jong-un ordered eight Musudan tests between mid-April and mid-October 2016. Seven failed, some spectacularly, before he ordered a full suspension of the effort. An 88 percent failure rate was unheard-of, especially for a proven design. The Musudan was based on a compact but long-range missile the Soviets had built in the 1960s for launching from submarines. Its small size but high power made it perfect for Kim's new strategy: shipping missiles around the country on mobile launchers and storing them in mountain tunnels, where American satellites would have trouble finding them.

As part of his effort to boost the range and lethality of the North's missile fleet, Kim had invested heavily in modifying the Soviet engines. The Musudan was far more complex than the Scud, the short-range missiles the North had made billions selling to Egypt, Pakistan, Syria, Libya, and Yemen, among other nations. Developing the Musudan technology was critical for Kim: He hoped it would pave the way for a whole new generation of single-stage and multistage missiles. With those in his arsenal, he could make good on his threat that no American base in the Pacific—and ultimately no American city—would be beyond his reach.

The North had been in the missile-launching business for a long time and had gained a reputation for mastering the art. So the serial run of Musudan failures in 2016—three in April, two in May and June, then two more in October, after the North had taken a pause to figure out what was happening to them—was confounding. The history of missile testing suggested that everyone suffered a lot of failures in the beginning, then figured it out and made things work. That's what happened during the race between the United States and the Soviet Union to build intercontinental ballistic missiles in the 1950s and '60s, an era marked by many spectacular crashes before the engineers and missileers figured out the technology. The Musudan experience reversed the usual trend. After years of successful tests of other missiles, it was as if North Korea's engineers forgot everything they'd learned.

Kim and his scientists were highly aware of what the United States and Israel had done to the Iranian nuclear program, and they had tried to insulate themselves from the same kind of attack. But the high failure rate of the missiles forced the North Korean leader to reassess the possibility that someone—maybe the Americans, maybe the South Koreans—was sabotaging his system. By October 2016, reports emerged that Kim Jong-un had ordered an investigation into whether the United States had somehow incapacitated the electronic guts of the missiles, perhaps getting inside their electronics or their command-and-control systems. And there was always the possibility that an insider was involved, or even several.

After each North Korean failure, the Pentagon would announce that it had detected a test, and frequently would even celebrate the missile's failure. "It was a fiery, catastrophic attempt at a launch that was unsuccessful," a Pentagon spokesman told reporters in April 2016, after the first full test of the Musudan, timed to celebrate the birthday of the country's founder, Kim Il-sung. When subsequent attempts failed, the official news release from the Pentagon included dryly worded boilerplate that "The North American Aerospace Defense Command determined the missile launch from North Korea did not pose a threat to North America." The statements never speculated about what went wrong.

But there was a lot of speculation inside the Pentagon, the NSA, and the White House, among the select group who knew about the classified US program to escalate cyber and electronic attacks against North Korea, with a particular focus on its missile tests. Each explosion, each case of a missile going off course and falling into the sea, prompted the same urgent question: "Was this because of us?"

It had been more than two years since Obama, alarmed by North Korea's progress, had pressed the Pentagon in early 2014 to drastically accelerate the effort to bring down North Korea's missiles—and turned again to cyber and electronic sabotage for the solution to geopolitical tensions. A lot had happened since then. The Sony attack had focused the administration's attention on North Korea, but on its cyberattacks, not its missile program. The negotiations with Iran—which led to a deal in the summer of 2015 that shipped 97 percent of Iran's nuclear fuel out of the country, setting back its efforts by a decade or more—consumed the attention of Washington's nuclear experts. Russia emerged as a far greater aggressor, and China demonstrated, with surprising vigor, that it was in search of influence, economic dominance, and a military presence in places it had never before ventured. The emergence of Donald

Trump made for captivating television as he transformed from a late-show punch line to an unstoppable candidate.

Through it all, Obama's North Korea sabotage effort churned ahead, silently.

The Obama administration's hope, of course, was that after two years of figuring out how to get inside North Korea's missile program, the United States had developed a worthy successor to Olympic Games: a way to delay by several years the day when the North would be able to threaten American cities with nuclear weapons. "It's too late to roll back the nuclear weapons program itself," William Perry, the former secretary of defense, told me. "Disrupting their tests would be a pretty effective way of stopping their ICBM program." It was the US government's only hope. The public strategy—which the White House briefly called "strategic patience"—was a failure. No diplomacy was under way. A military strike was far too risky. That left only covert action. And in countering proliferation, as one veteran of the process said to me wearily, "the best you can do is buy time."

The American-led cyber and electronic attacks on North Korea's missile program were vastly more complicated than the plan to go after the underground centrifuges in Iran years before. The Natanz nuclear plant was a comparatively easy target: It was a fixed site in a highly wired society, where engineers, diplomats, business executives, and scholars flowed in and out—all potential candidates to bring the malware into the country. And as the NSA and the Mossad were writing code to destroy the underground centrifuges, they had the luxury of time. As one veteran cyber warrior noted, if you got the code wrong, you could take it back to the shop, tinker with it, and try again in a week, a month, or six months. If the centrifuges then spun up or down too quickly and destroyed themselves, it was a pretty good bet the code had worked.

Going after North Korea's missiles was a completely different challenge. Access was miserable. The missiles were fired from multiple sites around the country, and increasingly from mobile launchers, in an elaborate shell game that was intended to mask the time and location of launches. And timing was everything. There was a tiny window for action to interfere with a launch: just as the missile was being fueled and prepared for liftoff, or in the seconds just after liftoff.

Even as North Korea's missiles exploded or fell into the sea, it was maddeningly difficult to understand exactly why. What proportion, if any, of the North's troubles arose from Obama's initiative? And what proportion was from other causes? Indeed, throughout the Pentagon and at the NSA and Cyber Command, the project targeting North Korea's missile program had created a lot of skeptics who doubted that a cousin of Olympic Games explained the North's troubles.

With every Musudan launch, the raw data about how the missiles performed—speed, trajectory, engine performance—were picked up by American early warning satellites and radar. The information then flowed back to the Pacific Command in Hawaii, to the Strategic Command in Omaha, and to Adm. Rogers's teams at Cyber Command and the NSA. The data were then picked apart by the Korea hands and weapons-of-mass-destruction experts at the CIA and fed into the computers of the Defense Intelligence Agency's Missile and Space Intelligence Center in Huntsville, Alabama. "I bet NASA didn't spend as much time breaking down moon launches as we've spent looking at Kim's missile tests," one American official later told me.

But in the end, no one could convincingly determine whether the program Obama had ordered was working. When the missiles flew or shattered, they took with them the best evidence of their precise condition at the time of failure. Centrifuges slowed, but missiles vanished. The teams of cyber and electronic experts who had been targeting the North Korean systems for years would show up at the Pentagon and draw a direct line from the cyber and electronic warfare program to Kim Jong-un's rocket troubles. Clearly, they had a strong interest in making that case: They wanted to show results from the huge, secret American investment in cyberweapons, at least in part to secure funding for new initiatives at Cyber Command. But according to several officials, they could never prove that any individual launch failed because of American interference.

Then the missile analysts would arrive, with alternative explanations. Yes, they conceded, the high rate of failure could have been accelerated because of the all-hands-on-deck effort to find ways into North Korea's systems. But there was no way to tell for sure. There were other possible explanations. The failures could have happened because of bad parts—especially because the United States and its allies had been running programs for more than a decade to get inside the North's supply chain. Or they could have happened because the North Korean engineers weren't as smart as they thought they were. Or maybe they were welding the rocket casings wrong.

"You have to be cautious whenever the enthusiasts of cyberattacks come in and claim victory," one former official advised me.

Whatever the true reason, the American plan to throw the missiles off-course succeeded at one thing: It made Kim Jong-un and his quartet of missile builders paranoid. The four members of the leadership who showed up in photographs surrounding the young leader during launches were clearly wondering whether sabotage, incompetence, or a series of unlucky accidents was the source of their misfortune. In that regard, the cyber sabotage effort initiated by the United States triggered the same kind of anxieties in North Korea that Stuxnet had caused in Iran, where the centrifuges seemed to be spinning normally—until unexplained disaster struck. The psychological effects may have been as important as the physical effects.

Nonetheless, the notoriously volatile young North Korean leader—known for executing his uncle and mounting a nerve-gas attack that killed his half-brother—proved remarkably tolerant when it came to the shortcomings of his rocket team. "We have never heard of him killing scientists," said Choi Hyun-kyoo, a researcher in South Korea who runs NK Tech, which manages a database of North Korean scientific publications. "He is someone who understands that trial and error are part of doing science."

Kim's missile whisperers could only hope his leniency would continue.

BEFORE THE NORTH KOREAN missiles began blowing up, I remembered only vaguely hearing the term "left of launch."

I knew the basics: that "left of launch" meant working to stop missiles *before* they were fired, when they are presumably easier to target. The phrase had an echo from the war in Iraq, where the military often used the shorthand "left of boom" to describe their effort to find and dismantle roadside bombs before they did damage.

But as a matter of international law and geopolitics, "left of launch" was far more fraught. At its core was the idea that the United States was prepared to mount a strike against another nation in peacetime, getting inside their infrastructure to attack their missile and command-and-control systems before they could be used against the United States. Of course, if a president ordered such a strike in the traditional way—say, by sending bombers in to destroy a missile base in peacetime—it would likely trigger a war. The hope was that by turning to cyberweapons or other sabotage, the United States could slip in far more subtly, deny responsibility for whatever happened, and get away without being caught.

Not surprisingly, on the rare occasions when the Pentagon talked about "left of launch" in public—and it did not happen often—they made it sound far more benign. They never used the word "preemption," knowing that word would raise a host of legal and political problems, starting with the obvious one: that only Congress can declare wars. Officials never even described "left of launch" as one of the president's options for covert action, something he could initiate by signing a presidential "finding" authorizing the intelligence agencies to take action in the defense of the United States.

Instead, "left of launch" was treated simply as another form of missile defense, a way to improve the chances of success for more traditional missile defenses—the antimissile systems that were supposed to hit an incoming nuclear warhead before it reached American shores.

Those traditional systems needed all the help they could get. The United States began working on antimissile defenses after the Soviet Union test-fired the world's first ICBM in 1957. That launch spurred President Dwight D. Eisenhower to initiate a crash program that swept in many of the nation's best scientists. Sixty years and more than $300 billion later, the concept of traditional missile-defense systems hadn't changed much. The aim was still to "hit a bullet with a bullet"—in other words, to intercept a warhead in midflight with a precision-guided antimissile system launched, often into space, from Alaska, California, or a ship at sea.

Given the number of Soviet missiles that could be launched at once, any American system would be overwhelmed. Later, after the Soviet Union fell, President George W. Bush focused on North Korea, which at the time could only hope to lob a few missiles in the direction of the United States. In late 2002, Bush announced that his administration was deploying long-range antimissile interceptors at a giant, muddy base just south of Fairbanks with a sister installation in California.

Once again, optimism outran experience. The number of successful interceptions in trial runs was embarrassing—roughly 50 percent, and that rate of success was achieved in tests conducted under ideal conditions. Soon the Pentagon stopped making public any quantifiable measurements. The truth was just too painful. Whenever senators pushed for more details, someone would tell them they would be happy to take it up in a classified session.

In light of these disappointments, at the Pentagon, and even among the defense contractors who were dependent on multibillion-dollar contracts for traditional missile interceptors, "left of launch" grew in importance. If missiles could be stopped on the ground, or in the first few seconds of flight, the interceptors would not have to be launched at all—they would become a backup defense, instead of a primary defense. The savviest of the contractors, eager to bid on the new business, began talking about "missile defeat" programs instead of "missile defense" programs. But the biggest of those contractors quietly worried that if cyber and electronic methods of taking down missiles proved too successful, they could put their multibillion-dollar traditional antimissile programs out of business. The big money was still in bending metal and making interceptors—not in writing code.

"This stuff is a double-edged sword," one person advising a major defense contractor told me. "Everyone wants it to work. But they don't want it to work too well."

WASHINGTON HID MANY of the details of the effort in plain sight.

When Bill Broad and I began asking around the Pentagon and the White House in 2016 about the surprising number of North Korean missile failures, we were not surprised to be met with stony silences. After the Stuxnet leak investigation, no one wanted to be accused of talking about a cyber-sabotage program—especially one that might not be working. But there were occasional hints that the answers lay in the "left of launch" program.

Bill dug into the open literature. Soon he showed up with a grin at my desk in Washington toting an inch-and-a-half-thick pile of Pentagon testimony and public documents. For a secret program, he noted, people sure had said a lot, mostly because they were lobbying for money.

The trail started with Gen. Martin E. Dempsey, chairman of the Joint Chiefs of Staff when Obama was pushing for the stepped-up attacks. Shortly after the North Koreans set off a nuclear test in February 2013, Dempsey publicly announced a new "left of launch" effort that would focus on "cyberwarfare, directed energy, and electronic attack." It was part of a larger presentation at the Pentagon on technologies that needed to be in place over the next seven years, and almost no one noticed the "left of launch" piece. But the plain fact was that the nation's top military officer had explained that malware, lasers, and signal jamming were all becoming important new adjuncts to the traditional methods of halting potential enemy strikes.

General Dempsey never mentioned North Korea in his statement. He didn't have to. A map accompanying the policy paper the Pentagon issued on the subject showed a missile from North Korea streaking toward the United States. That freed others to use similar imagery.

Soon Raytheon, the largest missile-defense contractor, started talking openly at conferences about the new opportunities in "left of launch" technologies, particularly cyber and electronic strikes executed at the moment of launch. A Raytheon document from one of its industry conferences, which was posted on a public website until we began asking questions about it, was not exactly subtle. One slide showed a range of adversaries against whom "left of launch" was particularly well suited, with a picture of a solemn Kim Jong-un sandwiched between Vladimir Putin and Xi Jinping. A chart illustrating how the program worked featured a bright band separating the steps that Raytheon's technology could accomplish to defeat missiles before and after launch. The most interesting part was the band itself—representing the minutes right around launch. There, Raytheon had inserted the words "cyber" and "EW," or electronic warfare, indicating that was the time to strike the launch process, at its most vulnerable point.

The chart showed that the cyber and electronic strikes could also target enemy factories—the latest effort at using industrial sabotage to slow the North. The program required a huge and complex effort, involving America's national laboratories, the Energy Department, and the CIA, and it was deployed against Iran as well. But it was hardly a surefire approach. The North Koreans were learning how to build more and more of their systems indigenously, and were even beginning to make some of the highly volatile rocket fuel that would power their longest-range missiles.

That progress made it all the more urgent that the National Security Agency and its Tailored Access Operations unit get inside the North's systems. Naturally, those operations remain among the most classified. But a tiny glimpse of the effort came from Oren J. Falkowitz, a quirky former NSA operator who started an innovative Silicon

Valley cybersecurity firm named Area 1. In a *Times* interview about his start-up, with my colleague Nicole Perlroth, he described how some of the company's approaches to anticipating cyberattacks were inspired by work done inside the NSA to break into computer systems for the North's missile program, in what he characterized as an effort to understand their missile-launch schedules.

Falkowitz said nothing about what the United States did with the information it acquired, which left open the question of whether we were just conducting espionage about launch schedules or were actively seeding implants in the North Korean systems. But there were other indications—some public, some whispered—of American successes in getting into the command-and-control systems. It was hard gaining access to the North's sealed-off computer networks, former American and South Korean operators reported. But once inside, they said, the North's digital defenses fell pretty quickly. North Korea's military, one noted, was as paranoid as Iran's, but not as talented.

A review of a gathering of top antimissile experts at the Center for Strategic and International Studies in 2015 gave us even more details. Archer Macy Jr., a retired navy rear admiral, described how the Pentagon was developing ways not only of preventing successful missile launches but also of interfering in their flight paths and navigation systems. That was followed by congressional testimony in which James Syring, the director of the Pentagon's Missile Defense Agency, described "left of launch" strikes as "game changing" because they reduced the need to "rely exclusively on expensive interceptors."

Every once in a while during these conferences and hearings, someone would touch on the profound question at the core of the program. That happened one day when Kenneth Todorov, a retired air force brigadier general, asked how the United States would justify what amounts to preemptive war under international law: attacking the North's missile launches first, before any strike, to gain a strategic advantage. "Are we, as a military and a nation," he asked, prepared to "go after potential targets in advance?" And if so, are we ready for other nations to do the same to us?

Todorov was getting at a critical point that has been periodically debated since President Bush, in 2002, declared that preemption was back as a central American principle for dealing with a hostile world. If the United States saw a missile on a North Korean launch pad being fueled, loaded with a warhead, and seemingly intended for American territory or that of an ally, it would likely be within its rights under international law to take out the missile on the pad.

But "left of launch" suggested a different scenario: A *preventive* strike, the kind that one state executes against another in the absence of an imminent threat. Think Pearl Harbor, or of a strong state that strikes a weaker but rising competitor while it still can. That is largely forbidden by international law.

With cyberstrikes—invisible, deniable—the temptation to conduct preventive war may be higher than it has ever been before. Unsurprisingly, few government officials want to delve too deeply, at least in public, into how the laws of war apply to offensive cyber action.

In private they debate these issues constantly. But as Robert Litt, the former general counsel to the director of national intelligence during the Obama years, put it to me one day: "There is no issue on which government lawyers have spent more time, to less productive effect, than on the question of how the laws of war apply to cyber."

IN MARCH 2016, just as North Korea was stepping up testing of its prized Musudan missile, I tried to engage Donald Trump on the subject of what he thought about America's new cyber arsenal—and how he might use it if elected. Trump was at Mar-a-Lago, his Florida golf club, and my colleague Maggie Haberman and I were interviewing him as part of a detailed conversation he'd agreed to with the *Times* on the subject of national-security issues.

My goal in trying to get him to talk about cyberweapons in that interview was simple: I wanted to see whether a candidate who spoke about military power as if it were still 1959—tanks and aircraft carriers and nukes—had given any thought to new technologies. For anyone new to the world of diplomacy, coercion, and military planning, the first step would be to understand the newest tools in the toolbox.

Digital warfare was new stuff for him; as the conversation went on, it wasn't clear he had ever heard of the American cyber operations against Iran. His main interest was to demonstrate, on cyber and all other issues, that he would be tougher and more decisive than Barack Obama, even if he wasn't quite certain what Obama had done in the cyber arena. He made an argument that, as with so many other things in Trump's worldview, America was blowing its lead:

> We're the ones that sort of were very much involved with the creation, but we're so obsolete, we just seem to be toyed with by so many different countries, already. And we don't know who's doing what. We don't know who's got the power, who's got that capability, some people say it's China, some people say it's Russia. But certainly cyber has to be a, you know, certainly cyber has to be in our thought process, very strongly in our thought process. Inconceivable that, inconceivable the power of cyber. But as you say, you can take out, you can take out, you can make countries nonfunctioning with a strong use of cyber. I don't think we're there. I don't think we're as advanced as other countries are, and I think you probably would agree with that. I don't think we're advanced, I think we're going backwards in so many different ways. I think we're going backwards with our military . . . we move forward with cyber, but other countries are moving forward at a much more rapid pace.

It was more assertion than analysis, more declaration than doctrine. We were trying to get Trump to discuss when it is justifiable to use cyberweapons; he took the conversation to the question of who is stronger and who is weaker, unencumbered by many facts.

Then, just to reinforce his campaign talking points, he concluded with: "We are frankly not being led very well in terms of the protection of this country."

Ten months later, Trump was inaugurated as the forty-fifth president and inherited a complex cyber operation against a hostile state that he barely understood. In the meantime, Bill Broad and I had built a pretty compelling case that North Korea was the target of an intensive, sophisticated US effort to send its missiles awry. With more reporting, we arrived at some solid conclusions about how the attacks worked.

Then came the sensitive part: telling the government what we were preparing to publish, seeking their comments, and hearing them out if they believed any of our revelations could compromise an ongoing operation or put lives at risk. In the last weeks of the Obama administration, we met with intelligence officials, fully anticipating that their first reaction would be to tell us to refrain from printing a story on the sensitive subject. When they said nothing of the sort, we left the session thinking we still had more reporting to do.

That reporting ran into the chaos of Trump's inauguration and his tumultuous first month in office—the blitz of executive orders, Trump's growing paranoia about the Russia investigation, and his suspicion of a "deep state" out to undermine him and his agenda. We were not ready to publish until late February. That's when I called K. T. McFarland, then Trump's deputy national security adviser, and explained to her that I needed to come by and make sure the new administration was aware of a story about a major program they were inheriting.

The next day I showed up at McFarland's tiny West Wing office. Her boss, Lt. Gen. Michael Flynn, had just been fired days before for misleading Vice President Michael Pence about his conversations with the Russian ambassador to the United States, Sergey Kislyak. Flynn had denied talking to Kislyak about overturning election-related sanctions against Russia that Obama had imposed in the last weeks of his presidency; in fact, the topic had indeed come up during their conversations.

As I walked past the national security adviser's corner office, the door was open; someone had rolled up the carpet, stripped all the books off the shelves, and stacked Flynn's office chair on top of his desk. It looked like a dorm room on move-out day, and certainly not the sight one expected to see a little more than a month into a new administration. It was a symbol of far more chaos to come.

While I knew the Obama transition teams had left binders full of briefing materials on North Korea for the new administration, I suspected few people had the clearances—or the time—to go through them all. Flynn, the former director of the Defense Intelligence Agency, was probably the one most current on the North Korean threat. But not only had he just been fired, his handpicked aides—derisively called "The Flynnstones"—were gradually being eased out.

When I sat down with McFarland, who had been a junior aide to Henry Kissinger in the White House forty years before, she told me that the administration had taken seriously Barack Obama's warning that North Korea would be their most immediate national-security problem. McFarland's job—in which she didn't last very long—was to convene the "deputies committee"—made up of the second- and third-ranking officials in State, Defense, Energy, Treasury, and the intelligence agencies—to tee up strategies for the president and his cabinet. Unsurprisingly, most of the initial meetings were about North Korea.

But as I began to describe to McFarland what we had learned about the "left of launch" program, and how it was being used against North Korea, I could see by the look on her face that it seemed to be the first she had heard of it. That was surprising: If there was anything that the new national-security team needed to get up to speed on quickly, it was the full range of American efforts to defang the North Korean threat. Perhaps she was just a good poker player, but the discussion did not suggest the new administration had a full grasp of what it was about to face.

After a half hour, McFarland said she had to go brief the president on a different matter. But she told me she saw no national-security issues in what we were planning to publish.

"It sounds like it will all work out," she said as she headed to the Oval Office.

She proved prematurely optimistic. After the word spread inside the administration, which had no experience dealing with sensitive national-security stories, the *Times* got a stiffly worded letter from the White House counsel, Donald McGahn, previously an election lawyer, accusing us of preparing to violate American national security—and hinting that the government might try to take some kind of action.

Within a few days Flynn's replacement as national security adviser, H. R. McMaster, invited us into his office to hear for himself what we were preparing to publish. It was his first full day on the job. A strategist with a PhD in military history, and the author of an incisive history of how the American military lied to itself about the war in Vietnam, his mind went straight to historical analogy.

"Is this the Enigma codes?" he asked, a reference to the encrypted German communications that the British cracked—a secret kept for decades. Broad and I told him we didn't think so: there was solid evidence that Kim Jong-un already understood the issue, and he had already shut down the Musudan tests after the string of failures.

McMaster had not had to deal with Korea or with cyber issues indepth before; he made his name in the Persian Gulf and had been promoted through the ranks by Gen. David Petraeus. His most recent jobs had been running the Army Capabilities Integration Center, where he was tasked with thinking about future conflicts. But he was clearly still catching up on the scope of the Korea crisis.

He asked us to meet yet again with intelligence officials and talk through the details. A day later we descended into the Situation Room to review our findings with them. Based on previous discussions, we had already decided to omit technical details, including several that might give the North indications of where their systems were vulnerable. That was normal practice. But explaining what the United States was doing, we thought, was vital: As our executive editor, Dean Baquet, noted, there was no way for Americans to have an informed public discussion about the US response to the North Korea crisis without understanding our past struggles to deal with it. "This was one of America's most urgent threats," he said, and that meant covering the American use of cyberweapons "the way we covered the Pentagon Papers, WikiLeaks, drone strikes, counterterrorism, and nuclear arms."

With publication imminent, McMaster went to brief Trump on what we were revealing—and what we were withholding. Trump had already ramped up his critique of the *Times,* and McMaster cautioned me that Trump might well take to Twitter to denounce the paper— hardly a first. In fact, on the morning that the story was published, Trump started a Twitter attack. But it wasn't about us.

"How low has President Obama gone to tapp [*sic*] my phones during the very sacred election process," he wrote that morning. "This is Nixon/Watergate." It was an accusation based on no facts.

Trump's tweet crystallized how his obsessions, and the chaos of the transition in the first six weeks of the new presidency, had prevented the new administration from focusing on what Obama had warned was the central national-security threat the nation faced. They had been left hundreds of pages of briefing materials about North Korea, but it appears little of it was absorbed. The questions swirling around the success or failure of the primary covert program to thwart the missile launches had not been fully engaged by McFarland, who was ousted in a few weeks, and they were entirely new to McMaster, who lasted just a bit more than a year.

Clearly Trump's mind was not yet on the dictator he would soon call "Rocket Man," or the country he would threaten to incinerate with "fire and fury." But positions were beginning to harden. Nineteen days before Trump's inauguration, Kim Jong-un had taunted the president-elect with a declaration that he was then in "the final stage in preparations" for an inaugural test of his intercontinental ballistic missiles—missiles larger and more sophisticated than the Musudan. Trump had responded, with typical Twitter bravado, "It won't happen."

It seemed inevitable that Trump would soon face the same challenge his predecessors did: how to deal with North Korea without prompting a broader war. He would confront issues that had been long debated in the Situation Room: whether to order the escalation of the Pentagon's cyber- and electronic-warfare effort, crack down again on trade with crushing economic sanctions, open negotiations with the North to freeze its nuclear and missile programs, or prepare for direct missile strikes on its nuclear and missile sites.

It seemed clear to me that, still lacking a strategy, Trump's answer would likely be to attempt all four.

WHILE THE UNITED STATES struggled to sabotage Kim's missile program, the North's hackers were looking for new targets in the West. In the two years after the Sony attack, their cyber corps had learned a lot and grown more global. As a top cybersecurity official for one of the behemoths of Silicon Valley put it to me, "If there was a 'most improved' award for states looking to weaponize the Internet, the North Koreans would win it. Hands down."

While Americans were thinking about how to use cyberweapons to neutralize the North's missiles, the North was thinking about how to use them to pay for those missiles—a huge challenge for a country under every form of economic sanction. Which is how the North's hacking teams cooked up a plan to steal $1 billion from the Bangladesh Central Bank in 2016.

With their exquisite nose for vulnerable institutions, the North's hackers focused on Bangladesh in January, figuring their cyber protections had to be pretty minimal. It was a good bet. With just a few weeks of quiet digital observation of the bank, the hackers got all they needed: the procedures for transferring funds internationally, some stolen credentials, and an understanding of when the bank would be closed for a holiday that extended into a weekend. The extra days provided them with time to execute transfers before anyone was around to stop them.

The hackers put together transfer orders for just under $1 billion, including one transfer to the Shalika Foundation in Sri Lanka. That proved the fatal mistake: In instructions to the New York Federal Reserve, through which such transactions flow, someone spelled "foundation" as "fandation." The error raised eyebrows, and the transfers were suspended—but only after Kim Jong-un's hackers had gotten away with $81 million. If it had been a physical bank heist, it would have been considered one of the largest and most brilliant in modern times. (By comparison, the great Brinks heist of 1950, in Boston's North End, swept up only about $2.7 million, worth about ten times that in modern currency.)

After the Sony hacks, the North had good reason to believe that any retaliation for their cyber exploits would be minimal, and they were right. There was no penalty for the Bangladesh bank attack, or cryptocurrency heists that followed.

"Cyber is a tailor-made instrument of power for them," Chris Inglis, a former deputy director of the National Security Agency, told me. "There's a low cost of entry, it's largely asymmetrical, there's some degree of anonymity and stealth in its use. It can hold large swaths of nation-state infrastructure and private-sector infrastructure at risk. It's a source of income."

At an earlier time, North Korea counterfeited crude $100 bills to finance the country's operations. That grew more difficult as the United States made the currency harder and harder to copy. But ransomware, digital bank heists, and hacks of South Korea's fledgling Bitcoin exchanges all made up for the loss of the counterfeiting business. Today the North may be the first state to use cybercrime to finance its state operations.

Bangladesh was hardly the only victim, and not even the first. In 2015 there was an intrusion into the Philippines, then the Tien Phong Bank in Vietnam. In February 2016 hackers got inside the website of Poland's financial regulator and infected visitors—from the central banks of Venezuela, Estonia, Chile, Brazil, and Mexico—in hopes of also breaking into those banks.

Then came two of the boldest attacks—one on South Korea, the other on the world.

There was no military document that the North wanted to read more than the American blueprints for war on the Korean Peninsula. Sometime in the fall of 2016, when most of the world was distracted by the presidential election, the North breached South Korea's Defense Integrated Data Center, according to Rhee Cheol-hee, a member of the South Korean parliament's National Defense Committee, and swept up 182 gigabytes of data—including OpPlan 5015, a detailed outline of what the US military delicately called a "decapitation strike." Rarely have the details leaked. But the documents the North's hackers stole appear to include strategies for finding and killing the country's top civilian and military leaders, and then wiping out as much of the mobile missile fleet and seizing as many nuclear weapons as possible. OpPlan 5015 would not stop there—the strategy included ways to counter the North's elite commandos, who would almost certainly slip into the South.

There is some speculation that the North intended to get caught stealing the war plan, in order to unnerve their adversaries and force them to rewrite it from scratch. We'll likely never know. But the theft is just one more sign of how deeply the North has compromised South Korea's sensitive networks. There is also evidence that Pyongyang has planted "digital sleeper cells" in critical infrastructure in the South in case they are needed to paralyze power supplies or command-and-control systems.

Then came WannaCry.

It is unclear how long the North Korean hacking team spent planning what the United States later charged was an "indiscriminate" attack on hundreds of thousands of computers, many in hospitals and schools. But it is clear how the hackers got inside: with some vulnerabilities in Microsoft software stolen from the NSA by the Shadow Brokers group. It was the ultimate cascading crime: the NSA lost its weapons; the North Koreans shot them back.

In this case, the hacking tool stolen from the NSA went by the name "Eternal Blue." It was a standard piece of the TAO's toolbox because it exploited a vulnerability in Microsoft Windows servers—an operating system so widely used that it allowed the malware to spread across millions of computer networks. No one had seen anything like it in nearly a decade, since a computer worm called "Conficker" went wild.

In this case, the North Korean hackers married the NSA's tool to a new form of ransomware, which locks computers and makes their data inaccessible—unless the user pays for an electronic key. The attack was spread via a basic phishing email, similar to the one used by Russian hackers in the attacks on the Democratic National Committee and other targets in 2016. It contained an encrypted, compressed file that evaded most virus-detection software. And once it burst alive inside a computer or network, users received a demand for $300 to unlock their data. It is unclear how many paid, but those who did never got the key—if there ever was one—to unlock their documents and databases.

The hackers guessed correctly that while Microsoft had patched this hole in the system—after the NSA had warned the company about the vulnerability just two months before the attack—few people who used old Microsoft Windows systems would have gone to the trouble of updating their software. And when the attackers struck in the late afternoon of May 12, 2016, anybody with ancient computers and ancient software to match—like the National Health Service in the United Kingdom—was a sitting duck.

"Many of the computers that were the most adversely affected were running Windows XP," Brad Smith, the president of Microsoft, explained to me later. "It's an operating system that we released in 2001. And when you stop and think about it, you realize that was six years before the first iPhone. It was six months before the first iPod." Smith didn't use the other obvious historic marker: the operating system was released to manufacturers just eighteen days before the September 11 attacks, a moment that changed our national sensibility about our vulnerabilities.

WannaCry, like the Russian attacks on the Ukraine power grid in the previous two years, was among a new generation of attacks that put civilians in the crosshairs. In that regard, it is akin to terrorism. "If you are wondering why you're getting hacked—or attempted-hacked—with greater frequency," said Jared Cohen, the former State Department official who now runs Alphabet's Jigsaw, a part of the Google parent company, which has done pioneering work in how to make people safer on the Internet, "it is because you are getting hit with the digital equivalent of shrapnel in an escalating state-against-state war, way out there in cyberspace."

Cohen is right: WannaCry is a prime example of where the newest cyber battles are headed. In the first years of state-on-state cyberwars, the targets of crippling hacks were mostly strategic, and often state-owned. Olympic Games was aimed at an isolated, underground nuclear enrichment facility. The attacks on ISIS were directed at vicious terrorist groups. The North Korea missile hacks were aimed at a program that directly threatened America and its allies.

But with WannaCry, the targeting seemed far more random, and the results were unpredictable. With computer systems of several major British hospital systems shut down, ambulances were diverted and non-emergency surgeries delayed. Banks and transportation systems across dozens of countries were affected. But it is doubtful the North Koreans knew, or cared, which systems would be crippled.

"I suspect the attackers had no idea what would be hit," one American investigator told me. "It was about creating chaos" and fear. Evidence of the untargeted nature of the malware lies in the fact that it hit seventy-four countries; after Britain, the hardest hit was Russia. (In what some might see as a sign of cosmic digital justice, Russia's Interior Ministry was among the most prominent victims.) Then Ukraine. Then Taiwan. There was no discernible political pattern.

Moreover, there was no warning. Britain's National Cyber Security Centre saw nothing coming, its director of operations, Paul Chichester, told my *Times* colleagues. In fact, investigators in Britain suspect the WannaCry attack may have been an early misfire of a weapon that was still under development—or a test of tactics and vulnerabilities.

"This was part of an evolving effort to find ways to disable key industries," said Brian Lord, a former deputy director for intelligence and cyber operations at Britain's GCHQ. "All I have to do is create a moderately disabling attack on a key part of the social infrastructure, and then watch the media sensationalize it and panic the public."

For all the billions spent on cyber defenses, in the end the Cyber Security Centre, British intelligence, and Microsoft had little to do with bringing the attack to an end. For that they had to thank Marcus Hutchins, a college dropout and self-taught hacker who was living with his parents in the southwest of England. He spotted a web address somewhere in the software and, largely on a lark, paid $10.69 to register it as a domain name as the attack was under way. The activation of the domain name turned out to act as a kill switch; it kept the malware from continuing to spread. (Hutchins was later arrested in Las Vegas and charged with being the author of another kind of malware, one designed to steal banking credentials.)

It took months—until December 2017, three years to the day after Obama accused North Korea of the Sony attacks—for the United States and Britain to formally declare that Kim Jong-un's government was responsible for WannaCry. Thomas Bossert, President Trump's homeland security adviser, said he was "comfortable" asserting that the hackers were "directed by the government of North Korea," but said that conclusion came from looking at "not only the operational infrastructure, but also the tradecraft and the routine and the behaviors that we've seen demonstrated in past attacks. And so you have to apply some gumshoe work here, not just some code analysis."

Bossert was honest about the fact that having identified the North Koreans, he couldn't do much else to them. "President Trump has used just about every lever you can use, short of starving the people of North Korea to death, to change their behavior," Bossert acknowledged. "And so we don't have a lot of room left here."

The gumshoe work stopped short, of course, of reporting about how Shadow Brokers allowed the North Koreans to get their hands on tools developed for the American cyber arsenal. Describing how the NSA enabled North Korean hackers was either too sensitive, too embarrassing, or both. And it was one of the most troubling parts of the whole incident.

While the US government says that it reports to industry more than 90 percent of the software flaws it discovers, so that they can be fixed, "Eternal Blue" was clearly part of the 10 percent it held on to in order to bolster American firepower. Microsoft never heard about the vulnerability until after the weapon based on it was stolen. Yet the US government acted as if it bore no responsibility for the devastating cyberattack. When I asked Bossert, and his deputy, Rob Joyce, who ran the TAO and clearly knew something of what happened to these pilfered weapons, they argued that the fault was entirely with those who used the weapons—not with those who lost control of them. It was a mystifying argument: if someone fails to lock up their guns, and a weapon stolen from their house is used in a school shooting, the gun owner has at least some moral or legal liability.

"It's a problem," Leon Panetta, the former defense secretary and CIA director told me one day as we discussed the WannaCry attacks, "when the US government can't hold on to its arsenal. We can't be in that position. And we wouldn't tolerate that explanation from other countries."

Brad Smith of Microsoft, clearly angry, compared the NSA's loss of its weapons to the air force's losing a Tomahawk Missile that was then shot back at an American ally. He pointed to the arrest of "an NSA contractor who had these weapons in his garage. And you don't see Tomahawk weapons in people's garages."

In fact, these days you did.

It was just two months later that Ukraine was hit with the Not-Petya attack, which roused Dymtro Shymkiv to action from upstate New York. It was very similar to WannaCry, although NotPetya was the work of the Russians, the Trump administration said in early 2017. Those hackers had clearly learned from the North Koreans. They made sure that no patch of Microsoft software would slow the spread of their code, and no "kill switch" could be activated.

In short, they designed a more accurate weapon, and struck two thousand targets around the world, in more than sixty-five countries. Maersk, the Danish shipping company, was among the worst hit: they reported losing $300 million in revenues and had to replace four thousand servers and thousands of computers. NotPetya made the Sony strike, only three years earlier, look like the work of amateurs.

WHATEVER THE CAUSE of Kim Jong-un's missile troubles in 2016—sabotage or incompetence or bad parts or faulty assembly—he solved the problem in 2017.

At a speed that caught American intelligence officials off guard—to say nothing of the newly arrived Trump administration—Kim rolled out an entirely new missile technology. Clearly he had a parallel program running alongside the Musudan, and it was based on another decades-old Soviet engine design that powered intercontinental ballistic missiles.

Unlike the Musudan, this one worked, and it worked right away. In quick succession Kim demonstrated ranges that could reach Guam, then the West Coast, then Chicago and Washington, DC. Out of nine intermediate and long-range launch tests in 2017, only one failed. That was an 88 percent success rate—a startling improvement from the year before.

And on the first Sunday in September, Kim detonated a sixth nuclear bomb, one that was far more powerful than any the North had set off before. It was fifteen times greater in power than the atomic bomb that leveled Hiroshima. Kim had entered the big leagues of nuclear power.

Many have seen this coming. For twenty years public CIA estimates declared that North Korea would have this capability sometime before 2020, but Kim's burst of progress after such a string of failures the previous year had not been predicted. Like the Russia hacks during the US election, Kim's strategic move caught the intelligence world unawares.

I went back to see General McMaster in December 2017. He readily acknowledged that Kim's race to the finish line—a bid to establish the North as a nuclear power before any negotiations began or sanctions took a more punishing toll—"has been quicker and the timeline is a lot more compressed than most people believed."

The question he and other officials would not touch, of course, was whether the North's string of successes in 2017 indicated that they had figured out the vulnerabilities of the Musudan—and solved them. What happened to "left of launch"? Were the new missiles less vulnerable to cyber and electronic attacks? Or had the supply chain changed, making it harder to infiltrate bad parts into the missile program? Or had the United States concluded it was simply being too obvious in attacking the Musudan and now was holding back until it was ready to strike at a larger missile?

There were plenty of indications that the US reliance on cyber tools was alive and well, just somewhat better hidden. Trump asked Congress in November 2017 for $4 billion in emergency funds for boosting missile defense and taking other steps to contain the North. Hundreds of millions of dollars were dedicated to what the budget documents called "disruption/defeat" efforts. Those efforts, several officials confirmed, include a more sophisticated attempt at cyber and electronic strikes. And there were several billion dollars allotted for traditional missile defense—even amid the doubts that it will work.

Trump's former CIA director, Mike Pompeo, occasionally hinted at ongoing programs, suggesting that the United States was "working diligently" to slow Kim's progress and delay the day when he was ready to put a nuclear warhead atop one of those missiles. Pompeo suggested that day was just "months away," but he repeated this estimate from early in Trump's administration through its first eighteen months. Jim Mattis, the defense secretary, had a darker take: after the North's most successful missile test, in November 2017, he said the country already had the ability to hit "everywhere in the world, basically."

AFTERWORD

SEN. DAN SULLIVAN (R-ALASKA): What do you think our adversaries think right now? If you do a cyberattack on America, what's going to happen to them?

LT. GEN. PAUL NAKASONE (COMMANDER OF US ARMY CYBER COMMAND): So basically, I would say right now they do not think that much will happen to them.

SULLIVAN: They don't fear us.

NAKASONE: They don't fear us.

SULLIVAN: So is that good?

NAKASONE: It is not good, Senator.

—*Lt. Gen. Paul Nakasone's confirmation hearing,*
as commander of US Cyber Command, March 1, 2018

UNTIL THE CYBER age came along, America's two oceans symbolized our enduring national myth of invulnerability. The threat of nuclear attack preoccupied us during the Cold War, but generally the United States has assured it could take out dictators, conduct drone strikes on terrorists, and blow up missile bases in faraway lands with relatively little fear of retaliation. There were exceptions, of course, moments of national terror: the British burned Washington in the War of 1812, the Japanese attacked Pearl Harbor, and al Qaeda brought down the Twin Towers and struck the Pentagon. But we knew the only attack that could threaten the existence of the country would come at the tip of a Soviet or Chinese intercontinental missile, or in the form of terrorists with access to nuclear weapons. And after some terrifying close calls, notably the Cuban Missile Crisis in 1962, we found an uneasy balance of power with our primary adversaries—mutually assured destruction—to deter the worst. It worked, or has so far, because the cost of failure is so high.

In the cyber age, we have not found that balance, and probably never will. Cyberweapons are entirely different from nuclear arms, and their effects have so far remained relatively modest. But to assume that will continue to be true is to assume we understand the destructive power of the technology we have unleashed and that we can manage it. History suggests that is a risky bet.

I keep on my desk a wonderful volume, *Airships in Peace and War*, first published in London in 1908 by military historian R. P. Hearne, which tried to imagine how a strange new invention of that time— airplanes—would change the course of history for Europe's great powers. One chapter is entitled "Could England Be Raided?" The question was answered in 1916, when the Germans first delivered scattered air attacks across the country. Within a year the first battles for control of the skies were under way. In 1940, the Blitz devastated London.

In the cyber world, we have not yet seen the equivalent of the Blitz. The early damage has been limited—centrifuges in Iran, a steel plant in Germany, a casino in Las Vegas, a crippled petrochemical plant in Saudi Arabia, missiles gone mysteriously awry in North Korea. Yet every week seems to bring hints of things to come, as city services became paralyzed by ransomware in Atlanta and patients were turned away after a cyberattack struck the health-care system in Britain.

The sheer acceleration in the number of attacks, and their rapidly changing goals, is one of several warning signs that we all are living through a revolution, playing out at digital speed.

IN THE EARLY days of this revolution, reaching for a cyberweapon seemed almost risk-free. Now that calculus is changing.

No one could blame an American president for using a remote-control weapon to crash Iran's nuclear centrifuges or disable North Korea's missiles. Given the choice between risking the lives of American soldiers or intelligence officers and reaching deep inside a country without setting foot in its territory, the decision seemed self-evident. The same logic that made drones so appealing to George W. Bush and Barack Obama—great stealth and low risk—made cyberweapons irresistible too. And in both Iran and North Korea, cyberweapons provided a way to slow dangerous military programs without triggering a war.

The harder question over the next decade will be whether reaching for such weapons with increasing frequency will continue to be a wise choice. By going into the North's missile systems, the United States set a precedent, just as we did with Olympic Games, that other nations will surely follow. While we talk publicly about setting norms for what should be off-limits for offensive cyber activity—hospitals, emergency responders, and now election systems—we are seen around the world as hypocrites. Every time the United States reaches into another nation's critical infrastructure, we make our own fair game for retaliation.

Yet we clearly are not prepared for the day when each American action in cyberspace triggers an escalating response. Because for now, as the stories told in the preceding pages make clear, deterrence is not working in the cyber realm. True, there has not been a devastating attack on the power grid, a "cyber Pearl Harbor" that might tempt an American president to make good on the threat contained in the 2018 Nuclear Posture Review, which is that some kinds of non-nuclear attacks—chiefly, cyberattacks—may force the president to reach for the ultimate weapon.

The very fact that we need to make the threat underscores the failures of the past few years. When Adm. Michael Rogers took over the National Security Agency, he told me in his office in 2014 that his tenure would be measured by his success at convincing America's adversaries that there was a cost—a high cost—to attacking the nation's networks. "Right now, if you look at most nation-states—groups and individuals and the activity they are engaging in in cyber, very broadly, most of them seem to have come to the conclusion that there is little risk of having to pay a price for this in real terms," he said at Stanford later that year.

When his successor, General Nakasone, conceded in his confirmation hearing four years later that "they don't fear us," he was admitting that after spending billions of dollars on new defenses and new offensive weapons, the United States has still failed to create a deterrent against cyberattacks.

Perhaps that is understandable. In the Cold War, nuclear deterrence did not emerge instantly. It took years of collaboration between technologists, strategists, generals, and politicians. It involved a very public debate, which the United States seems unwilling to conduct in the cyber realm—for fear of revealing our capabilities, or having to surrender some of them.

In the nuclear era, deterrence worked well between the United States and the Soviet Union not only because each knew the other possessed world-destroying power, but also because each had confidence in the integrity of its own weapons system. Each was certain that if the president ordered a launch, the launch would happen.

But over the past few years we have seen time and again that cyberweapons can undermine that confidence. The Iranians lost all assurance they could control their centrifuges. The North Koreans suspected someone was messing with their launch systems. And inside the Pentagon there is growing fear that one day in the not-too-distant future an American commander could order a launch and missiles would not fire.

We experienced a less deadly version of that loss of confidence in 2016, when we feared that Russian hackers were seeking to break into our election systems, looking for ways to alter voter-registration data. Even if they failed, the mere attempt was enough to undercut public confidence in the outcome of the vote. Imagine a similarly skilled group breaking into America's nuclear early-warning systems, triggering a fake warning that the United States was under attack. It could prompt a president to launch our own weapons before the chimerical incoming missiles could strike.

This may sound like the stuff of a bad thriller, but almost exactly that scenario—without the cyber manipulation—nearly triggered disaster in 1979, when a watch officer awoke William Perry, then an undersecretary of defense, to report that an early-warning system was showing two hundred incoming ICBMs. The military quickly determined it to be a false alarm: someone had placed a training tape, simulating an incoming attack, into the real warning system. However, Perry later warned, if an enemy attempted the same thing with a sophisticated bit of malware, perhaps placed by an insider, "we might not be so lucky next time."

The implications of having our own command-and-control system compromised underscore why sabotaging similar systems in other nations is dangerous business. If American leaders—or Russian leaders—feared their missiles might not lift off when someone hit the button, or that they were programmed to go off-course, it could easily undermine the system of deterrence that has helped reduce the likelihood of nuclear war for the past several decades. It could also encourage countries to build more missiles—as an insurance policy—and perhaps to launch them earlier.

"It's not hard to imagine how we greatly increase the risk of stumbling into a conflict because of an accident, or inadvertence, or just deliberate deception," James Miller, a former undersecretary of defense for policy, and one of the country's most experienced nuclear strategists, told me after he and Richard Fontaine finished a study of just that problem. "It's conceivable that other states, and even non-state actors, could undertake cyberattacks that lead to an inadvertent escalation with Russia," Miller concluded. That a president could make snap decisions on which millions of lives depend, based on information that had been subtly manipulated, is sheer madness.

General Nakasone's warning that countries do not fear us—one he uttered just weeks before becoming the new director of the NSA and commander of United States Cyber Command—focused on the question of whether the United States can retaliate after its networks are struck. But there are other ways to deter attacks—chiefly by convincing your adversaries that your defenses are strong, and they will not succeed. In the lingo of strategists, this is called "deterrence by denial." If an attack would be futile, why bother in the first place?

Deterrence by denial requires an exquisite defense. And while American intelligence officials will not concede the point, internal government assessments say it will be a decade—at least—before the United States can reasonably defend our most critical infrastructure from a devastating cyberattack launched by Russia or China, the two most skilled adversaries in the field. There are simply too many vital networks, growing too quickly, to mount a convincing defense. Offense is still wildly outpacing defense. As Bruce Schneier, a cyber expert whose work is a must-read on the topic, put it so well: "We are getting better. But we are getting worse faster."

Schneier's point is that even as we build far greater defenses, our vulnerabilities are expanding dramatically. With huge investments, the top tier of the financial industry and the electric utilities have done the best job of safeguarding their networks—meaning that a North Korean hacker aiming at those industries would likely have more luck targeting smaller banks and rural power companies. But as we put autonomous cars on the road, connect Alexas to our lights and our thermostats, put ill-protected Internet-connected video cameras on our houses, and conduct our financial lives over our cell phones, our vulnerabilities expand exponentially.

During the Cold War, we learned how to live, uneasily, with the knowledge that the Soviet Union and China had nuclear weapons pointed at us. There were no perfect defenses. In a world of constant cyber conflict we will have to adjust similarly.

Yet if we are more vulnerable than ever, why is the Pentagon talking about the need to conduct a far more aggressive cyber strategy? In testimony to Congress in early 2018, the leaders of the NSA and Cyber Command pressed the case that if the United States is to prevail in the new era of cyber conflict, our forces need to be unshackled. Even if we see attacks massing, they said, the current rules of engagement keep us from attacking the attackers. It is time, they argued, to start "hacking the hackers."

The approach Cyber Command described and detailed in its strategy documents is one of nearly daily raids behind enemy lines, looking for threats before they reach America's own computer networks. "The United States must increase resiliency, defend forward as close as possible to the origin of adversary activity, and persistently contest malicious cyberspace actors to generate continuous tactical, operational, and strategic advantage," one of those documents said—all military-speak for taking the war to the enemy.

It was an instinct born of more than a decade of counterterrorism operations, where the United States learned that the best way to take on al Qaeda or ISIS was by destroying them at their bases and in their living rooms. But in cyber it amounts to an admission that our defenses at home are wildly insufficient and that the only way to win is to respond to every perceived threat. As with many of Trump's new strategies, taken to its logical extreme this approach carries enormous risks of miscalculation and escalation. To pull it off, the United States would have to scrap the requirement that the president authorize every destructive cyberattack. Cyber operations would begin to look more like evening raids conducted by Special Operations Forces. The problem is that when other countries adopt the same strategy, as inevitably they will, the chances rise dramatically that cyberattacks will accelerate and could trigger a shooting war, or worse.

. . .

So what is to be done?

The first step is to recognize the folly of going on offense unless we have a good defense. We would be lucky to seal up three-quarters of the glaring vulnerabilities in American networks today. But the best way to deter attack—and counterattack—is deterrence by denial. That requires a major national effort, far beyond the civil defense projects of the 1950s when the United States built a highway system that could evacuate civilians and dug shelters in large cities. A parallel effort to secure America's cyber infrastructure has often been discussed, but it has never happened. It is complicated by the fact that the main targets of attack are in private hands. Given the complexity of the Internet, the government can't regulate how banks, telecom firms, gas pipeline companies, and Google and Facebook design their cybersecurity. Every one of those systems is radically different.

For that reason, even after a decade of debate it's still not clear who in the federal government, if anyone, is responsible for defending the country—and the economy—from the most sophisticated cyber-attacks. Homeland Security is supposed to "coordinate," but just as we expect the Pentagon to defend the United States against incoming missile attacks, there's a presumption that it will defend American companies and individuals against sophisticated, state-sponsored hacks (but not against scammers, teenage hackers, and trolls living in Saint Petersburg). It's time to get real. The government isn't going to play a role in protecting American institutions except when it comes to the most critical of infrastructures: the electric grid, the voting system, the water and wastewater systems, the financial system, and nuclear weapons. Once we've understood this fact, we need a Manhattan Project to lock down our most critical systems. That will take presidential leadership.

Even then, civil defense will not be close to enough. One of the lessons of the past few years is that the dynamic of cyberattacks is completely different from what we grew accustomed to during the superpower standoffs of the twentieth century. We have to adjust our strategy to reflect that we will be far more vulnerable than almost any other major nation for years to come. As Michael Sulmeyer, a former Pentagon official now running a Harvard cyber initiative, has observed, "When it comes to cyberspace . . . the United States has more to lose than its adversaries because it has gone further in embracing innovation and connectivity without security. But although the societies and infrastructure of Washington's adversaries are less connected and vulnerable, their methods of hacking can still be disrupted. . . .

"If the United States hopes to win," he continued, "it should spend less time trying to persuade its competitors that it is not worth hacking and more time preempting them and degrading their ability to do so. It is time to target capabilities not calculations."

What does that mean in the real world? Obviously, the United States is not going to respond to every cyberattack; we would be in constant low-level war. Not every cyberattack needs a cyber response. Criminal attacks should be handled as other crimes are handled—with vigorous prosecution. The United States is getting better and better at that: the indictments of Iranian and Chinese hackers—even if they are still at large—and the extradition of a major Russian cyber criminal in 2018 show there are ways of responding short of treating every hack as if it is an attack.

And as in everything else in global affairs, red lines matter. So when trolls from the Internet Research Agency began bombarding the United States with fake news from fake accounts—with the intent of meddling in an American election—they needed to be delisted from Facebook. (That happened, but not until well after the election.) If the agency remained undeterred, its servers needed to be melted down, courtesy of our cyberweapons. The servers would be replaced, of course, perhaps quickly. But the message would be sent, and the Russians would know that the United States was able and willing to respond.

And while the intelligence agencies would insist on secrecy, that would defeat the point: for our response to deter attackers, it needs to be very public—as public as an American airstrike on a chemical-weapons plant in Syria, or an Israeli strike on a nuclear reactor. Every time we respond quietly—or not at all—to an attack because we are worried about revealing the quality of our detection systems or the capability of our weapons, we only encourage escalation and further cyber strikes from our adversaries.

For the same reason, the United States needs to open up about some of our own offensive cyber operations, especially if their details have been revealed. To this day the United States has not admitted its role in Olympic Games. It was, after all, a covert operation—and covert operations are not to be discussed, by law. But what if, once the code was traveling around the world and it became widely known that Stuxnet was an American-Israeli creation, both Washington and Jerusalem had publicly owned up to their role? What if they had admitted to it, the way Israel acknowledges, implicitly or explicitly, that it has bombed reactors in Iraq and Syria? We might well have established one of those red lines: if you produce nuclear fuel in violation of UN mandates, expect that something bad could happen to your centrifuges—maybe from the air, maybe from cyberspace.

Most important, just as the United States must show other nations there is a price to pay for truly serious cyberattacks, we must also show that some things are off-limits. And until America discusses publicly—at the presidential level—what we *will not* do in cyberspace, we have no hope of getting other countries to limit themselves as well.

IT WILL BE easier to navigate those decisions when the government acknowledges a few realities.

The first is that our cyber capabilities are no longer unique. Russia and China have nearly matched America's cyber skills; Iran and North Korea will likely do so soon, if they haven't already. We have to adjust to that reality. Those countries will no sooner abandon their cyber arsenals than they will abandon their nuclear arsenals or ambitions. The clock cannot be turned back. So it is time for arms control.

Second, we need a playbook for responding to attacks, and we need to demonstrate a willingness to use it. It is one thing to convene a "Cyber Action Group," as Obama did fairly often, and have them debate when there is enough evidence and enough concern to recommend to the president a "proportional response." It is another thing to respond quickly and effectively when such an attack occurs.

Third, we must develop our abilities to attribute attacks and make calling out any adversary the standard response to cyber aggression. The Trump administration, in its first eighteen months, began doing just this: it named North Korea as the culprit in WannaCry and Russia as the creator of NotPetya. It needs to do that more often, and faster.

Fourth, we need to rethink the wisdom of reflexive secrecy around our cyber capabilities. Certainly, some secrecy about how our cyberweapons work is necessary—though by now, after Snowden and Shadow Brokers, there is not much mystery left. America's adversaries have a pretty complete picture of how the United States breaks into the darkest corners of cyberspace.

But the intelligence agency's insistence on secrecy—the refusal to discuss offensive cyberweapons in any detail—makes it impossible to debate how precisely these weapons can be targeted and whether some should be banned because of their potential threat to civilians. We cannot expect Russian and Iranian hackers to stop implanting malware in our utility grid unless we are willing to talk about giving up our own implants in their power grids. We cannot insist that the US government has the right to a "backdoor" into Apple's iPhones and encrypted apps unless we are willing to make the Internet less safe for everyone, because any backdoor will become the target of hackers around the globe.

No country likes giving up military or intelligence capabilities. But we have done it before. America swore off chemical and biological weapons when we determined that the cost to civilians of legitimizing them was greater than any military advantage they offered. We limited the kinds of nuclear weapons we would build, and banned some. We can do the same in cyberspace, but only if we are willing to openly discuss our capabilities and to help monitor cyberspace for violators.

Fifth, the world needs to move ahead with setting these norms of behavior even if governments are not yet ready. Classic arms-control treaties won't work: they take years to negotiate and more to ratify. With the blistering pace of technological change in cyber, they would be outdated before they ever went into effect. The best hope is to reach a consensus on principles that begins with minimizing the danger to ordinary civilians, the fundamental political goal of most rules of warfare. There are several ways to accomplish that goal, all of them with significant drawbacks. But the most intriguing, to my mind, has emerged under the rubric of a "Digital Geneva Convention," in which companies—not countries—take the lead in the short term. But countries must then step up their games too.

Microsoft's president, Brad Smith, is one of the strongest advocates of the concept. He imagines loosely modeling a cyber accord among companies on traditional warfare conventions that have evolved for more than a century. Over the decades the rules have broadened and deepened, embracing the treatment of prisoners, the banning of chemical weapons, the protection of noncombatants, and the kind of aid that should be provided to the wounded, no matter whose side they fought on.

The analogy to cyberspace is hardly exact. The Geneva Conventions apply in wartime; if there is hope for an analogous set of rules of the road in cyber, they will need to set standards for peacetime. And they must apply to companies as well as countries, since Google, Microsoft, Facebook, and Cisco form the battlespace in which the world's cyber conflicts are fought.

In the spring of 2018, about three dozen companies—Microsoft, Facebook, and Intel among them—agreed to the most basic set of principles, including an innocent-sounding vow that the signatories would refuse to help any government, including the United States, mount cyberattacks against "innocent civilians and enterprises from anywhere." The companies also committed to come to the aid of any nation on the receiving end of such attacks, whether the motive for the attack is "criminal or geopolitical."

It was a start, but a barely satisfying one. No Chinese, Russian, or Iranian companies were part of the initial compact, nor were some of the biggest forces in the technology world, including Google and Amazon, both still struggling between their desires to do vast business with the US military and their desires to avoid alienating their customers. The wording of the accord left lots of maneuvering room for the companies to join attacks against terror groups, or even against governments repressing their own citizens. Moreover, the principles made no mention of supporting democracy, or human rights—meaning that Apple, if it later joined the accord, could still get away with its decision to bow to Beijing by keeping its data on Chinese customers on servers inside China. In other words, the first principles were like the Internet—sprawling and messy.

"I have no illusions this will be easy," Smith told me in Germany at the beginning of 2018. "We're going to need laws passed that make clear that certain principles need to be respected around the world, that governments need to refrain from attacking critical infrastructure in times of peace or war, or even when it's unclear whether we're at a time of peace or war." Of course, the Geneva Conventions have been regularly violated, in world wars and civil wars, from Vietnam to Syria.

There's no such thing as fully protecting civilians. Individual citizens don't have the option of going on the offense, and most have no interest in becoming combatants in a global cyber conflict. But over time, these principles have made the world more humane.

Still, there are steps individuals should take to protect themselves and help to avoid becoming collateral damage. Awareness—about what phishing campaigns look like, about how to lock up home-network wi-fi routers, and about how to sign up for two-factor authentication—can help to wipe out 80 percent or so of the daily threat. If we wouldn't leave our doors unlocked when we leave home, or the keys in the ignition of our cars, we shouldn't leave our lives exposed on our phones, either.

None of that will stop a determined, state-sponsored adversary. Houses can be protected against everyday burglars, but not against incoming ICBMs.

The lesson of the past decade is that, unless shooting breaks out, it will always be unclear if we are at peace or war. Governments that cannot stand up to far larger powers with conventional armies will have little incentive to give up the advantages that cyberweapons offer. We are living in a gray zone, one of constant digital conflict. That is not a pleasant prospect, but it is the world we have created for ourselves. To survive it, we must make some fundamental decisions, akin to ones we made after the invention of the airplane and the atomic bomb—decisions that enabled us to navigate a constant state of peril.

Now, as then, we have to think more broadly about where our security will be found. Clearly, it is not in an unending cyber arms race where victories over adversaries are fleeting, and where the greatest objective is to break another nation's encryption or turn off its factories. We need to remember that we built these technologies to enrich our societies and our lives, and not to find yet another way to plunge our adversaries into darkness. The good news is that because we created the technology, we have a chance of controlling it—if we concentrate on how to manage the risks. It has worked in other realms. It can work in cyberspace as well.

ACKNOWLEDGMENTS

T HE PERFECT WEAPON grew out of my reporting for *The New York Times*, but it is also a follow-on to a world I began to explore in *Confront and Conceal* (Crown, 2012). That book was the first to tell the story of Olympic Games, the American-Israeli cyber effort aimed at Iran's nuclear program. At the time it was published, it was hard to find more than a handful of examples of cases in which states used cyber-weapons against each other. Scarcely six years later, that is a daily oc-currence. So, not surprisingly, the ambitions for a book that explained this era grew, and with it so did my indebtedness to editors, researchers, and colleagues.

Let me start at the *Times*, where I have worked for nearly thirty-six years, in Washington and overseas. Arthur Sulzberger Jr. and A. G. Sulz-berger, our previous and current publishers, have been unstintingly generous in letting me roam the world to explain to our readers this new and frightening age. And they never complained about the legal bills. Dean Baquet, our executive editor, and Joe Kahn, the managing editor, have championed these stories, and pressed for more. So have Matt Purdy, Susan Chira, and Rebecca Corbett, who offered ideas, fine editing, and encouragement along the way.

In Washington, Elisabeth Bumiller, the Washington bureau chief, a relentless champion of investigative efforts and a friend since our days in Japan, a quarter century ago, allowed me the free rein to report and the leave to take some time to write the book. Bill Hamilton, an extraordinary national security editor, made every story he touched far better. My thanks as well to Lara Jakes, Amy Fiscus, and Thom Shanker, editors who pressed for more facts, better sources, and clearer explanations.

The daily miracle of the Washington bureau of the *Times* is the reporting staff, and I have been lucky enough to join forces on many of these stories with colleagues and friends. Eric Lipton and Scott Shane knew it was the moment to tell a bigger story about the Russia investigation in the fall of 2016, and together we produced a lengthy reconstruction of the Russia hack from which the title of this book is borrowed. That story was among the entries for the 2017 Pulitzer Prize in International Reporting, won with a group of *Times* reporters around the world who delved deeply into Vladimir Putin's information-warfare techniques. I am indebted to that entire team, whose reporting enriched my understanding of the Russia story.

In Washington, Silicon Valley, and abroad my reporting colleagues Eric Schmitt, Mark Landler, Mark Mazzetti, Peter Baker, Matthew Rosenberg, Matt Apuzzo, Julie Davis, Nicole Perlroth, David Kirkpatrick, Alison Smale, Steve Erlanger, Matt Apuzzo, and Adam Goldman all joined forces on the intersection of foreign policy, cyber, and law enforcement. Maggie Haberman and I teamed up during the presidential campaign for two lengthy interviews with Donald Trump that helped me understand his evolving views on national security—and gave me a chance to raise cyber issues that seemed entirely new to him.

A special thanks to my reporting partner of three decades, Bill Broad, who understood how cyber, nuclear, and missile technology issues converge—and every day brought his reporting skills and unerring instincts to the hardest stories, particularly the American effort to sabotage North Korea's missile program.

David McCraw, the *Times'* exceptional in-house lawyer, got me through the leak investigations surrounding Olympic Games and helped me around others, while offering expert advice about how to tell the story of America's activities in cyberspace.

Harvard's Belfer Center for Science and International Affairs at the Kennedy School of Government has long been my intellectual community for grappling with the strategic implications of cyber, and its scholars and former policy makers were generous with their time and their willingness to educate a journalist. I have had the privilege of co-teaching, with Graham Allison and Derek Reveron, "Central Challenges in American National Security, Strategy and the Press." Graham's legendary strategic insights, and the course's mix of graduate students, military and intelligence fellows, and undergraduates led to fascinating explorations of the complexity of cyber conflict. My special thanks to Joseph Nye, Ashton B. Carter, Eric Rosenbach, Michael Sulmeyer, R. Nicholas Burns, Rolf Mowatt-Larssen, and Ben Buchanan. Drew Faust, Harvard's president for the past decade, provided constant encouragement and asked me to test out my thoughts with a variety of audiences.

And when I needed a place in Washington to settle in and write, Jane Harman and Robert Litwak opened the doors of the Wilson Center for International Scholars, a remarkable institution of calm and deep thought in a capital that could use a lot more of both. I am grateful to both of them and to Meg King, who has devoted herself to making Wilson a place for Congress to learn about cyber.

This book could not have been written without the aid of a remarkable group of research assistants, drawn from our course at Harvard. The most critical of them has been Alyza Sebenius, an incredibly talented young reporter, writer, and editor. Alyza headed the team, conducting interviews, editing chapters, gracefully pushing me to dig deeper, write more clearly, and think about those readers for whom the subject matter may seem daunting. She investigated, drafted, reorganized, and kept the project going—and is an example of an inspiring generation that is making American journalism as vital, and vitally important, as at any moment in our history.

Mary Brooks devoted nights and weekends to understanding China's outsized role in cyber conflict, fact-checked and edited, and was indispensable to the process of turning stories into chapters, and chapters into arguments. Ana Moran delved into ISIS's activities on the web and Silicon Valley's involvement. Sohum Pawar thought deeply about the lessons of the Ukraine hack and guided us through complex technology, as did Anand Gupta. They started down this path hoping to learn something from me, but I learned far more from each of them—and they made this project possible. Gabrielle Chefitz and Josh Cohen provided helpful research support.

At Stanford, my thanks to Amy Zegart, Herb Lin, Phil Taubman, Michael McFaul, and Condoleezza Rice for counsel, ideas, and a base of operations when I was reporting in the technology world.

Michael Carlisle has been a friend for more than three decades and a remarkable book agent and counselor. He guided me to the Crown imprint at Penguin Random House, where I learned why Kevin Doughten is considered one of the finest editors in the business. It was Kevin who pressed for a book that would explore the geopolitics of this revolution, and it was his energy, insights into how to tell the story, fascination with the new technology, and willingness to work around the clock that made *The Perfect Weapon* possible. Jon Darga, Annsley Rosner, Rachel Rokicki, Penny Simon, Julie Cepler, Kathleen Quinlan, Courtney Snyder, Mark Birkey, Linnea Knollmueller, Kirsten Clawson, and Elizabeth Rendfleisch made Crown's magic happen. Amelia Zalcman provided stellar legal advice.

Molly Stern, Crown's publisher, has never flinched from hard topics, and she has been an enthusiastic advocate of telling this story. I am lucky to be in Crown's stable of writers.

Alex Gibney, Javier Botero, and Sarah Dowland, documentary-makers extraordinaire at Jigsaw Productions, had the inspiration that the story of Olympic Games told in *Confront and Conceal* should be a film, and in *Zero Days*, shown in theaters and on Showtime in 2016, they pushed the story forward. Some of their new reporting, especially on the operation Nitro Zeus, is represented in this book.

None of this—the reporting, the writing, the support—would be possible without my love, Sherill, the best editor and partner possible. Everything she touches in this world she makes better—and her editing skills saved us yet again. Andrew Sanger, our elder son and a recent graduate of Colorado College, delved into fact-checking and turned a critical eye to the explanations of history and technology; his brother Ned, a Harvard undergraduate, reviewed key chapters.

My parents, Ken and Joan Sanger, pushed me to get the best education possible, encouraged my start in journalism, and have been a source of support and love since, along with my sister Ellin and her husband, Mort Agress.

This is a work of current history, in a subject area in which far too much is classified. So, by definition, this account cannot be comprehensive; years from now we will learn about operations, internal disputes, successes and failures that are still cloaked. The best I can offer is that it represents the most accurate understanding of the incidents and debates I have been able to render. The errors of fact or interpretation are, of course, my own.

David E. Sanger
Washington, DC
May 2018

NOTES

PREFACE

xi **startling recommendation:** "Nuclear Posture Review," Office of the Secretary of Defense, February 2018, www.defense.gov/News/SpecialReports/2018 NuclearPostureReview.aspx.

xi **the recommendation leaked immediately:** David E. Sanger and William Broad, "Pentagon Suggests Countering Devastating Cyberattacks with Nuclear Arms," *New York Times,* January 17, 2018, www.nytimes.com/2018/01/16/us/politics/pentagon-nuclear-review-cyberattack-trump.html.

xii **Terrorism topped that list:** John D. Negroponte, "Annual Threat Assessment of the Director of National Intelligence," January 11, 2007, www.dni.gov/files/documents/Newsroom/Testimonies/20070111_testimony.pdf.

xii **"Great power competition":** Helene Cooper, "Military Shifts Focus to Threats by Russia and China, Not Terrorism," *New York Times,* January 20, 2018, www.nytimes.com/2018/01/19/us/politics/military-china-russia-terrorism-focus.html.

xiv **"so obsolete in cyber":** "Transcript: Donald Trump Expounds on His Foreign Policy Views," *New York Times,* March 26, 2016, www.nytimes.com/2016/03/27/us/politics/donald-trump-transcript.html.

xvii **"So much of the fabric of our society":** Joyce spoke at the Aspen Institute in Washington, DC, on November 15, 2017: www.aspeninstitute.org/events/cyber-breakfast-view-from-the-white-house/.

xviii **hybrid war:** Valery Gerasimov, "The Value of Science Is in the Foresight," *Military-Industrial Courier,* February 2013.

xix **"little price to pay"**: Andrew Desiderio, "NSA Boss Suggests Trump Lets Putin Think 'Little Price to Pay' for Messing With U.S.," *Daily Beast,* February 27, 2018, www.thedailybeast.com/nsa-boss-seems-to-hit-trump-on-russia -putin-believes-little-price-to-pay-for-messing-with-us.

xix **Wilbur and Orville:** Andrew Glass, "President Taft Witnesses Wright Brothers Flight, July 29, 1909," *Politico,* July 29, 2016, www.politico.com/story/ 2016/07/president-taft-witnesses-wright-brothers-flight-july-29-1909-226158.

xx **airplanes manufactured in the United States:** Stephen Budiansky, *Air Power* (New York: Penguin Books, 2004).

PROLOGUE: FROM RUSSIA, WITH LOVE

2 **Ukraine was a playground and testing ground:** *Wired*'s Andy Greenberg wrote some of the best pieces on the Ukraine hack. See "How an Entire Nation Became Russia's Test Lab for Cyberwar," June 20, 2017, www.wired.com/ story/russian-hackers-attack-ukraine/.

4 **sitting in darkness:** "Cyber-Attack Against Ukrainian Critical Infrastructure," Industrial Control Systems, Cyber Emergency Response Team, February 25, 2016, ics-cert.us-cert.gov/alerts/IR-ALERT-H-16-056-01.

5 **shut them off at will:** Nicole Perlroth and David E. Sanger, "Cyberattacks Put Russian Fingers on the Switch at Power Plants, U.S. Says," *New York Times,* March 16, 2018, www.nytimes.com/2018/03/15/us/politics/russia-cyberattacks .html.

CHAPTER I: ORIGINAL SINS

7 **"Old Headquarters"**: Steve Hendrix, "Former OSS Spies on a Mission to Save Old Headquarters," *Washington Post,* June 28, 2014, www.washingtonpost .com/local/former-oss-spies-on-a-mission-to-save-old-headquarters/2014/ 06/28/69379d16-fd7d-11e3-932c-0a55b81f48ce_story.html?utm_term =.0e1c8190b76b.

10 **As our story explained:** David E. Sanger, "Obama Order Sped Up Wave of Cyberattacks Against Iran," *New York Times,* June 1, 2012, www.nytimes .com/2012/06/01/world/middleeast/obama-ordered-wave-of-cyberattacks -against-iran.html.

11 **That is exactly what happened:** My story of Stuxnet and its political history was *Confront and Conceal: Obama's Secret Wars and Surprising Use of American Power* (New York: Crown Publishers, 2012), and excerpted in the *Times* on June 1, 2012. In a "Note on Sources" in the book, I wrote that "I discussed with senior government officials the potential risks of publication of sensitive information that touches on ongoing intelligence operations." But I gave no details of those discussions, which were off the record. In the intervening years, responding to Freedom of Information Act requests from other news organizations, the CIA has released emails indicating with whom I spoke and some of the content, while

redacting details it still regards as sensitive. Other details appeared in court filings surrounding the case of Gen. James A. Cartwright, who was charged with lying to the FBI in connection with the investigation into the book's revelations. (Cartwright was later pardoned by President Obama.) In providing this account, I am continuing to respect any off-the-record agreements we had surrounding the conversations for the book, except for material that has already been publicly released or that I have since been given permission to reveal.

12 **the FBI was called in to investigate:** Chris Doman, "The First Sophisticated Cyber Attacks": How Operation Moonlight Maze Made History," Medium, July 7, 2016, medium.com/@chris_doman/the-first-sophistiated-cyber-attacks -how-operation-moonlight-maze-made-history-2adb12cc43f7.

12 **Colorado School of Mines:** Ben Buchanan and Michael Sulmeyer, "Russia and Cyber Operations: Challenges and Opportunities for the Next U.S. Administration," Carnegie Endowment for International Peace, December 13, 2016, carnegieendowment.org/2016/12/13/russia-and-cyber-operations-challenges -and-opportunities-for-next-u.s.-administration-pub-66433.

13 **The hackers had lurked:** Doman, "The First Sophisticated Cyber Attacks."

13 **"This was a real wake-up call":** For a good summary of declassified documents on Moonlight Maze, see Ibid. Thomas Rid also has a helpful guide to the attacks in *Rise of the Machines: A Cybernetic History* (New York: W. W. Norton & Company, 2016).

14 **"targeted network":** Michael Hayden, *Playing to the Edge: American Intelligence in the Age of Terror* (New York: Penguin Books, 2016), 184.

17 **"tools available to a president":** Sam LaGrone, "Retired General Cartwright on the History of Cyber Warfare," *USNI News,* October 18, 2012, news.usni .org/2012/10/18/retired-general-cartwright-history-cyber-warfare.

18 **worldwide threat assessment:** John D. Negroponte, January 11, 2007, "Annual Threat Assessment of the Director of National Intelligence," www.dni .gov/files/documents/Newsroom/Testimonies/20070111_testimony.pdf.

18 **Chinese attacks on American companies:** Department of Justice, "Chinese National Who Conspired to Hack into U.S. Defense Contractors' Systems Sentenced to 46 Months in Federal Prison," July 13, 2016, www.justice.gov/ opa/pr/chinese-national-who-conspired-hack-us-defense-contractors-systems -sentenced-46-months.

18 **Lockheed Martin's networks:** Justin Ling, "Man Who Sold F-35 Secrets to China Pleads Guilty," VICE News, March 24, 2016, news.vice.com/article/ man-who-sold-f-35-secrets-to-china-pleads-guilty.

18 **Barack Obama and John McCain:** Lee Glendinning, "Obama, McCain Computers 'Hacked' During Election Campaign," *Guardian,* November 7, 2008, www.theguardian.com/global/2008/nov/07/obama-white-house-usa.

19 **the true wake-up call:** Ellen Nakashima, "Cyber Intruder Sparks Response,

Debate," *Washington Post,* December 6, 2011, www.washingtonpost.com/
national/national-security/cyber-intruder-sparks-response-debate/2011/12/
06/gIQAxLuFgO_story.html?utm_term=.ed05d5330dc5.

20 **The Russians had left USB drives:** William J. Lynn III, "Defending a New
Domain," *Foreign Affairs,* September–October 2010, www.foreignaffairs
.com/articles/united-states/2010-09-01/defending-new-domain.

21 **an equally important motivation:** Israeli defense minister Ehud Barak admitted as much to biographers in recorded interviews. Later, Israeli officials
asked how that news got past the country's military censors. Jodi Rudoren,
"Israel Came Close to Attacking Iran, Ex-Defense Minister Says," *New York
Times,* August 22, 2015, www.nytimes.com/2015/08/22/world/middleeast/
israel-came-close-to-attacking-iran-ex-defense-minister-says.html.

22 **President Bush had secretly authorized a covert plan:** David E. Sanger,
"U.S. Rejected Aid for Israeli Raid on Iranian Nuclear Site," *New York
Times,* January 10, 2009, www.nytimes.com/2009/01/11/washington/11iran
.html.

22 **the quiet engineer who dug in:** Together they became hacker heroes in *Zero
Days,* the 2016 Alex Gibney documentary, which was based in part on the
story about Stuxnet I told in *Confront and Conceal: Obama's Secret Wars and
Surprising Use of American Power* (New York: Crown Publishers, 2012). I have
benefited from much of the research that Gibney and his team conducted in
the two years they spent making the documentary, including the description
of "Nitro Zeus"—a plan to shut down Iran's power grid and other facilities
in case of war.

26 **spymaster Meir Dagan:** David E. Sanger, "A Spymaster Who Saw Cyberattacks as Israel's Best Weapon Against Iran," *New York Times,* March 23,
2016, www.nytimes.com/2016/03/23/world/middleeast/israel-mossad
-meir-dagan.html.

26 **"sticky bombs":** David E. Sanger, "America's Deadly Dynamics with Iran,"
New York Times, November 6, 2011, www.nytimes.com/2011/11/06/sunday
-review/the-secret-war-with-iran.html.

27 **On his desk:** Isabel Kershner, "Meir Dagan, Israeli Spymaster, Dies at 71;
Disrupted Iran's Nuclear Program," *New York Times,* March 18, 2016, www
.nytimes.com/2016/03/18/world/middleeast/meir-dagan-former-mossad
-director-dies-at-71.html.

27 **Dagan devoted his last years in office:** Ronen Bergman, *Rise and Kill First*
(New York: Random House, 2018), 623.

27 **"intolerable consequences":** Ronen Bergman, "When Israel Hatched a Secret
Plan to Assassinate Iranian Scientists," *Politico,* March 5, 2018, www.politico
.com/magazine/story/2018/03/05/israel-assassination-iranian-scientists
-217223.

28 **Iraq's Osirak nuclear reactor:** Mark Mazzetti and Helene Cooper, "U.S. Confirms Israeli Strikes Hit Syrian Target Last Week," *New York Times,* September 11, 2007, www.nytimes.com/2007/09/12/world/middleeast/12syria .html.

34 **"taken it to a new level":** Elad Benari, "McCain: Obama Leaked Info on Stuxnet Attack to Win Votes," *Israel National News,* April 6, 2012, www .israelnationalnews.com/News/News.aspx/156501.

34 **"we have mechanisms in place":** "Remarks by the President," June 8, 2012, James S. Brady Press Briefing Room, obamawhitehouse.archives.gov/the-press -office/2012/06/08/remarks-president.

CHAPTER II: PANDORA'S INBOX

37 **"The science-fiction cyberwar scenario is here":** Alex Gibney, dir., *Zero Days,* Magnolia Pictures, 2016.

38 **It formally came into existence:** Siobhan Gorman and Yochi Dreazen, "Military Command Is Created for Cyber Security," *Wall Street Journal,* June 24, 2009, www.wsj.com/articles/SB124579956278644449.

39 **"these are the kinds of [decisions] that are serious":** The full transcript of the interview with Carter, conducted at the Aspen Security Forum, is at archive .defense.gov/Transcripts/Transcript.aspx?TranscriptID=5277.

43 **Gates wrote a blistering memorandum:** I described this set of exchanges between Gates and Donilon in *The Inheritance* (New York: Crown Publishers, 2009), 185–86.

44 **"attack plan on this scale":** For the description of Nitro Zeus I am indebted to my friend Javier Botero, who headed an investigative team for Alex Gibney's production of *Zero Days,* a documentary about cyber conflict that was based in part on revelations in *Confront and Conceal.* Javier went far beyond my reporting, finding many former members of the military and civilian teams that engaged in the planning for the larger operation against Iran, and described its risks.

45 **the two secret cyber programs suggest:** David E. Sanger and Mark Mazzetti, "U.S. Had Cyberattack Plan If Iran Nuclear Dispute Led to Conflict," *New York Times,* February 17, 2017, www.nytimes.com/2016/02/17/world/middleeast/us -had-cyberattack-planned-if-iran-nuclear-negotiations-failed.html.

46 **"We have seen nation-states spending a lot of time and a lot of effort":** Damian Paletta, "NSA Chief Says Cyberattack at Pentagon Was Sophisticated, Persistent," *Wall Street Journal,* September 8, 2015, www.wsj.com/ articles/nsa-chief-says-cyberattack-at-pentagon-was-sophisticated-persistent -1441761541.

46 **"ask yourself why":** Ibid.

48 **"Don't pick on us"**: David E. Sanger, "U.S. Indicts 7 Iranians in Cyberattacks on Banks and a Dam," *New York Times,* March 25, 2016, www.nytimes.com/2016/03/25/world/middleeast/us-indicts-iranians-in-cyberattacks-on-banks-and-a-dam.html.

48 **announced the creation of cybercorps:** Thom Shanker and David E. Sanger, "U.S. Suspects Iran Was Behind a Wave of Cyberattacks," *New York Times,* October 14, 2012, www.nytimes.com/2012/10/14/world/middleeast/us-suspects-iranians-were-behind-a-wave-of-cyberattacks.html.

48 **Iranian hackers began targeting:** "Iranians Charged with Hacking U.S. Financial Sector," FBI, March 24, 2016, www.fbi.gov/news/stories/iranians-charged-with-hacking-us-financial-sector.

51 **Iranians struck Saudi Arabia:** Ross Colvin, " 'Cut Off Head of Snake' Saudis Told U.S. on Iran,' " Reuters, November 28, 2010, www.reuters.com/article/us-wikileaks-iran-saudis/cut-off-head-of-snake-saudis-told-u-s-on-iran-idUSTRE6AS02B20101129.

51 **Hackers found an easy target in Saudi Aramco:** Nicole Perlroth and David E. Sanger, "Cyberattacks Seem Meant to Destroy, Not Just Disrupt," *New York Times,* March 29, 2013, www.nytimes.com/2013/03/29/technology/corporate-cyberattackers-possibly-state-backed-now-seek-to-destroy-data.html.

51 **their hackers wreaked havoc:** David E. Sanger, David D. Kirkpatrick, and Nicole Perlroth, "The World Once Laughed at North Korean Cyberpower. No More," *New York Times,* October 16, 2017, www.nytimes.com/2017/10/15/world/asia/north-korea-hacking-cyber-sony.html.

52 **and the phones were dead:** This serves as more evidence, if needed, that companies that try to save money by putting their phones on the same "voice over Internet" networks that their computers run on are increasing their vulnerability.

52 **American intelligence agencies quickly concluded:** My colleague Nicole Perlroth did the best contemporaneous account of what happened at Saudi Aramco: "Cyberattack on Saudi Oil Firm Disquiets U.S.," *New York Times,* October 24, 2012, www.nytimes.com/2012/10/24/business/global/cyberattack-on-saudi-oil-firm-disquiets-us.html. CNN did a good reconstruction of the chaos at Saudi Aramco. See Jose Pagliery, "The Inside Story of the Biggest Hack in History," CNN, August 5, 2015, money.cnn.com/2015/08/05/technology/aramco-hack/index.html.

CHAPTER III: THE HUNDRED-DOLLAR TAKEDOWN

56 **"Someplace had to be last":** With my colleague Eric Schmitt, I described the role of the "crawler" in a story in the *New York Times* in early 2014. See David E. Sanger and Eric Schmitt, "Snowden Used Low-Cost Tool to Best NSA," February 9, 2014, www.nytimes.com/2014/02/09/us/snowden-used-low-cost-tool-to-best-nsa.html.

57 **a peek through the keyhole:** Some of my reporting for this chapter draws from a chapter I contributed to *Journalism After Snowden,* published by Columbia University Press in March 2017.

58 **The *Times* on the wrong side of his wrath:** David Sanger, "U.S. Rejected Aid for Israeli Raid on Iranian Nuclear Site," *New York Times,* January 10, 2009, www.nytimes.com/2009/01/11/washington/11iran.html.

58 **he wanted a job at the NSA:** Rachael King, "Ex-NSA Chief Details Snowden's Hiring at Agency, Booz Allen," *Wall Street Journal,* February 4, 2014, www .wsj.com/articles/exnsa-chief-details-snowden8217s-hiring-at-agency-booz -allen-1391569429.

60 **As one of my *Times* colleagues put it so well:** Scott Shane, "No Morsel Too Minuscule for All-Consuming N.S.A.," *New York Times,* November 3, 2013, www.nytimes.com/2013/11/03/world/no-morsel-too-minuscule-for-all -consuming-nsa.html.

60 **the "wild, wild West":** Barack Obama, in a speech given at Stanford in February 2015, obamawhitehouse.archives.gov/the-press-office/2015/02/13/remarks -president-cybersecurity-and-consumer-protection-summit.

61 **James Clapper said Snowden had taken advantage:** David Sanger and Eric Schmitt, "Spy Chief Says Snowden Took Advantage of 'Perfect Storm' of Security Lapses," *New York Times,* February 12, 2014, www.nytimes.com/2014/ 02/12/us/politics/spy-chief-says-snowden-took-advantage-of-perfect-storm-of -security-lapses.html?.

62 **But the NSA's solution was either too late:** Jo Becker, Adam Goldman, Michael S. Schmidt, and Matt Apuzzo, "N.S.A. Contractor Arrested in Possible New Theft of Secrets," *New York Times,* October 6, 2016, www.nytimes.com/ 2016/10/06/us/nsa-leak-booz-allen-hamilton.html.

63 **In the days after Snowden showed up:** David E. Sanger and Jeremy Peters, "A Promise of Changes for Access to Secrets," *New York Times,* June 14, 2013, www.nytimes.com/2013/06/14/us/nsa-chief-to-release-more-details-on -surveillance-programs.html?mtrref=www.google.com.

64 **That explains why about one-third:** Philip Bump, "America's Outsourced Spy Force, by the Numbers," *The Atlantic,* June 10, 2013, www.theatlantic.com/ national/archive/2013/06/contract-security-clearance-charts/314442/.

68 **In 2005 the air force hired the RAND Corporation:** Evan S. Medeiros et al., *A New Direction for China's Defense Industry,* Rand Corporation, 2005, www.rand.org/content/dam/rand/pubs/monographs/2005/RAND_MG334 .pdf.

68 **blocked the purchase:** Steven R. Weisman, "Sale of 3Com to Huawei Is Derailed by U.S. Security Concerns," *New York Times,* February 21, 2008, www.nytimes.com/2008/02/21/business/worldbusiness/21iht-3com .html.

69 **That was the name of a covert program:** David E. Sanger and Nicole Perlroth, "N.S.A. Breached Chinese Servers Seen as Security Threat," *New York Times,* March 23, 2014, www.nytimes.com/2014/03/23/world/asia/nsa-breached -chinese-servers-seen-as-spy-peril.html?.

72 **In late 2013, *Der Spiegel* published the "ANT catalog":** Spiegel Staff, Documents Reveal Top NSA Hacking Unit, December 29, 2013. www.spiegel.de/ international/world/the-nsa-uses-powerful-toolbox-in-effort-to-spy-on-global -networks-a-940969.html; Jacob Appelbaum, Judith Horchert, and Christian Stöcker, "Shopping for Spy Gear: Catalog Advertises NSA Toolbox," *Der Spiegel,* December 29, 2013, www.spiegel.de/international/world/catalog-reveals -nsa-has-back-doors-for-numerous-devices-a-940994.html.

73 **The catalog revealed:** I had been aware of these technologies in 2012, when I first published accounts of Olympic Games, in which they were important. But I withheld some of the details at the request of American officials who did not believe the Iranians yet understood how the technology worked. After the Snowden revelations, of course, they had a road map.

74 **"You have not heard me as the director say":** Maya Rhodan, "New NSA Chief: Snowden Didn't Do That Much Damage," *Time,* June 30, 2014, time .com/2940332/nsa-leaks-edward-snowden-michael-rogers/.

75 **But Merkel was outraged:** Alison Smale, "Germany, Too, Is Accused of Spying on Friends," *New York Times,* May 6, 2015, www.nytimes.com/2015/05/ 06/world/europe/scandal-over-spying-shakes-german-government.html.

75 **Unsatisfied, Merkel called Obama:** David E. Sanger and Alison Smale, "U.S.-Germany Intelligence Partnership Falters Over Spying," *New York Times,* December 17, 2013, www.nytimes.com/2013/12/17/world/europe/us-germany -intelligence-partnership-falters-over-spying.html?

76 **Still, intelligence leaders were unapologetic:** Mark Landler and Michael Schmidt, "Spying Known at Top Levels, Officials Say," *New York Times,* October 30, 2013, www.nytimes.com/2013/10/30/world/officials-say-white -house-knew-of-spying.html.

77 **"way beyond so-called domestic surveillance":** Eli Lake, "Spy Chief James Clapper: We Can't Stop Another Snowden," *Daily Beast,* February 23, 2014, www.thedailybeast.com/spy-chief-james-clapper-we-cant-stop-another -snowden.

CHAPTER IV: MAN IN THE MIDDLE

80 **When the *Washington Post* first published the slide:** Barton Gellman and Ashkan Soltani, "NSA Infiltrates Links to Yahoo, Google Data Centers Worldwide, Snowden Documents Say," *Washington Post,* October 30, 2013, www.washingtonpost.com/world/national-security/nsa-infiltrates-links-to -yahoo-google-data-centers-worldwide-snowden-documents-say/2013/10/30/ e51d661e-4166-11e3-8b74-d89d714ca4dd_story.html?

80 **"Fuck these guys"**: Brandon Downey, "This Is the Big Story in Tech Today," Google+ (blog), October 30, 2013, plus.google.com/+BrandonDowney/posts/ SfYy8xbDWGG.

84 **Google soon added a new email-encryption feature**: Ian Paul, "Google's Chrome Gmail Encryption Extension Hides NSA-Jabbing Easter Egg," *PC World,* June 5, 2014, www.pcworld.com/article/2360441/googles-chrome-email -encryption-extension-includes-jab-at-nsa.html.

85 **the existence of the NSA program**: Barton Gellman and Laura Poitras, "U.S., British Intelligence Mining Data from Nine U.S. Internet Companies in Broad Secret Program," *Washington Post,* June 7, 2016, www.washingtonpost .com/investigations/us-intelligence-mining-data-from-nine-us-internet -companies-in-broad-secret-program/2013/06/06/3a0c0da8-cebf-11e2-8845 -d970ccb04497_story.html.

86 **Mark Zuckerberg posted a heated defense**: Mark Zuckerberg, Facebook post, June 7, 2013, www.facebook.com/zuck/posts/10100828955847631.

86 **a long history of cooperation**: Julia Angwin, Charlie Savage, Jeff Larson, Henrik Moltke, Laura Poitras, James Risen, "AT&T Helped U.S. Spy on Internet on a Vast Scale," *New York Times,* August 16, 2015, www.nytimes.com/2015/ 08/16/us/politics/att-helped-nsa-spy-on-an-array-of-internet-traffic.html.

87 **Amazon's $600 million deal**: Aaron Gregg, "Amazon Launches New Cloud Storage Service for U.S. Spy Agencies," *Washington Post,* November 20, 2017, www.washingtonpost.com/news/business/wp/2017/11/20/amazon-launches -new-cloud-storage-service-for-u-s-spy-agencies/?utm_term=.8dcf7ac21a9f.

88 **Cook's social and political intuition**: Todd Frankel, "The Roots of Tim Cook's Activism Lie in Rural Alabama," *Washington Post,* March 7, 2016, www.washingtonpost.com/news/the-switch/wp/2016/03/07/in-rural-alabama -the-activist-roots-of-apples-tim-cook/?utm_term=.5f670fd2354d.

91 **"more than 5½ years"**: Computer-security experts question that figure, because Apple does not fully realize how quickly the NSA's supercomputers can crack codes.

92 **the agency developed the "Clipper chip"**: Steven Levy, "Battle of the Clipper Chip," *New York Times,* June 12, 1994, www.nytimes.com/1994/06/12/ magazine/battle-of-the-clipper-chip.html.

92 **the Clinton administration retreated**: Susan Landau, *Listening In: Cybersecurity in an Insecure Age* (New Haven: Yale University Press, 2017), 84.

92 **Morell and his colleagues sided with Big Tech**: Richard A. Clarke, Michael J. Morell, Geoffrey R. Stone, Cass Sunstein, Peter Swire, *Report and Recommendations of the President's Review Group on Intelligence and Communications Technologies,* December 12, 2013, lawfare.s3-us-west-2.amazonaws.com/ staging/s3fs-public/uploads/2013/12/Final-Report-RG.pdf.

93 **Comey predicted there would be a moment**: David E. Sanger and Brian

Chen, "Signaling Post-Snowden Era, New iPhone Locks Out N.S.A.," *New York Times,* September 27, 2014, www.nytimes.com/2014/09/27/technology/iphone-locks-out-the-nsa-signaling-a-post-snowden-era-.html.

95   **twenty-two were injured:** Adam Nagourney, Ian Lovett, and Richard Perez-Pena, "San Bernardino Shooting Kills at Least 14; Two Suspects Are Dead," *New York Times,* December 3, 2015, www.nytimes.com/2015/12/03/us/san-bernardino-shooting.html.

95   **three children with his wife:** "San Bernardino Shooting Victims: Who They Were," *Los Angeles Times,* December 17, 2015, www.latimes.com/local/lanow/la-me-ln-san-bernardino-shooting-victims-htmlstory.html.

97   **subsequent report by the FBI inspector general:** Office of the Inspector General, US Department of Justice, *A Special Inquiry Regarding the Accuracy of FBI Statements Concerning its Capabilities to Exploit an iPhone Seized During the San Bernardino Terror Attack Investigation,* March 2018, oig.justice.gov/reports/2018/o1803.pdf.

97   **He wrote a 1,100-word letter to his customers:** Eric Lichtblau and Katie Benner, "Apple Fights Order to Unlock San Bernardino Gunman's iPhone," *New York Times,* February 18, 2016, www.nytimes.com/2016/02/18/technology/apple-timothy-cook-fbi-san-bernardino.html.

97   **"The United States government":** A copy of this letter is viewable at www.apple.com/customer-letter/.

98   **FBI paid at least $1.3 million to a firm:** Eric Lichtblau and Katie Benner, "F.B.I. Director Suggests Bill for iPhone Hacking Topped $1.3 Million," *New York Times,* April 22, 2016, www.nytimes.com/2016/04/22/us/politics/fbi-director-suggests-bill-for-iphone-hacking-was-1-3-million.html.

98   **"If, technologically, it is possible":** Michael D. Shear, "Obama, at South by Southwest, Calls for Law Enforcement Access in Encryption Fight," *New York Times,* March 12, 2016, www.nytimes.com/2016/03/12/us/politics/obama-heads-to-south-by-southwest-festival-to-talk-about-technology.html.

CHAPTER V: THE CHINA RULES

100   **"there are two kinds of big companies":** Scott Pelley, "FBI Director on Threat of ISIS, Cybercrime," CBS News, October 5, 2014, www.cbsnews.com/news/fbi-director-james-comey-on-threat-of-isis-cybercrime/.

100   **a base for the People's Liberation Army:** I am indebted to two *Times* colleagues, David Barboza and Nicole Perlroth, with whom I worked on the *Times* investigation into Unit 61398. Some of the material from that original story is reproduced here, supplemented by subsequent reporting and the details in the US indictment of the Unit 61398 officers. David E. Sanger, David Barboza, and Nicole Perlroth, "Chinese Army Unit Is Seen as Tied to Hacking Against U.S.," *New York Times,* February 19, 2013, www.nytimes.com/2013/02/19/technology/chinas-army-is-seen-as-tied-to-hacking-against-us.html.

100 **from the designs of the F-35 aircraft:** David E. Sanger, "Chinese Curb Cyber-attacks on U.S. Interests, Report Finds," *New York Times,* June 21, 2016, www.nytimes.com/2016/06/21/us/politics/china-us-cyber-spying.html.

101 **attacks on 141 companies across nearly two dozen industries:** "APT1: Exposing One of China's Cyber Espionage Units," February 18, 2013, www.fireeye .com/content/dam/fireeye-www/services/pdfs/mandiant-apt1-report.pdf.

103 **"We know hackers steal people's identities":** "Remarks by the President in the State of the Union Address," White House Office of the Press Secretary, February 12, 2013, obamawhitehouse.archives.gov/the-press-office/2013/02/12/remarks-president-state-union-address.

104 **"Over the past four days, I have seen freedom":** Bill Clinton, "President Clinton's Beijing University Speech, 1998," US-China Institute, June 29, 1998, china.usc.edu/president-clintons-beijing-university-speech-1998.

105 **"God's gift to China":** Liu Xiaobo, "God's Gift to China," *Index on Censorship* 35, no. 4 (2006): 179–81.

105 **Bloomberg, among others, folded:** Edward Wong, "Bloomberg Code Keeps Articles from Chinese Eyes," *New York Times,* November 28, 2013, sinosphere.blogs .nytimes.com/2013/11/28/bloomberg-code-keeps-articles-from-chinese-eyes/.

106 **A secret State Department cable:** Described in the *New York Times* "State's Secrets" series in 2010. James Glanz and John Markoff, "Vast Hacking by a China Fearful of the Web," *New York Times,* December 5, 2010, www.nytimes .com/2010/12/05/world/asia/05wikileaks-china.html?pagewanted=print.

106 **"images of China's military":** Ibid.

106 **in December 2009, Google's top executives discovered:** David E. Sanger and John Markoff, "After Google's Stand on China, U.S. Treads Lightly," *New York Times,* January 15, 2010, www.nytimes.com/2010/01/15/world/asia/15diplo.html.

107 **"Operation Aurora":** Kim Zetter, "Google Hack Attack Was Ultra Sophisticated, New Details Show," *Wired,* January 14, 2010, www.wired.com/2010/01/operation-aurora/.

107 **Google took the bold step of announcing:** David Drummond, "A New Approach to China," Official Google Blog, January 12, 2010, googleblog.blogspot .com/2010/01/new-approach-to-china.html.

108 **"A well-placed contact claims":** Ellen Nakashima, "Chinese Leaders Ordered Google Hack, U.S. Cable Quotes Source as Saying," *Washington Post,* December 4, 2010, www.washingtonpost.com/wp-dyn/content/article/2010/12/04/AR2010120403323.html.

108 **"this may well mean having to shut down Google.cn":** Drummond, "A New Approach to China."

109 **"Knowing that you were subjects of an investigation":** Ellen Nakashima,

"Chinese Hackers Who Breached Google Gained Access to Sensitive Data, U.S. Officials Say," *Washington Post,* May 20, 2013, www.washingtonpost .com/world/national-security/chinese-hackers-who-breached-google-gained -access-to-sensitive-data-us-officials-say/2013/05/20/51330428-be34-11e2 -89c9-3be8095fe767_story.html?.

110 **Among the most colorful was a hacker:** Sanger, Barboza, and Perlroth, "Chinese Army Unit Is Seen as Tied to Hacking Against U.S."

110 **The target was a subsidiary of Telvent:** Ibid.

112 **Five million Americans:** Brian Fung, "5.1 Million Americans Have Security Clearances. That's More than the Entire Population of Norway," *Washington Post,* March 24, 2014, www.washingtonpost.com/news/the-switch/wp/2014/ 03/24/5-1-million-americans-have-security-clearances-thats-more-than-the -entire-population-of-norway/?utm_term=.88e88f78d45e.

112 **documented in a series of reports:** US House of Representatives, "The OPM Data Breach: How the Government Jeopardized Our National Security for More than a Generation," Committee on Oversight and Government Reform, September 7, 2016, oversight.house.gov/wp-content/uploads/2016/09/ The-OPM-Data-Breach-How-the-Government-Jeopardized-Our-National -Security-for-More-than-a-Generation.pdf.

113 **problems were so acute:** U.S. Office of Personnel Management Office of the Inspector General Office of Audits, "Federal Information Security Management Act Audit FY 2014," November 12, 2014, www.opm.gov/our-inspector -general/reports/2014/federal-information-security-management-act-audit -fy-2014-4a-ci-00-14-016.pdf.

113 **shutting down the system was not an option:** "Statement of the Honorable Katherine Archuleta," Hearing before the Senate Committee on Homeland Security and Governmental Affairs, June 25, 2015.

113 **Archuleta and her staff were clueless:** US House of Representatives, "The OPM Data Breach."

113 **The Chinese got caught once and expelled:** Ibid.

113 **At some point during the summer of 2014:** Ibid.

114 **a private computer-security contractor working for OPM flagged an error:** Brendan I. Koerner, "Inside the Cyberattack That Shocked the US Government," *Wired,* October 23, 2016, www.wired.com/2016/10/inside-cyberattack -shocked-us-government/.

114 **"They are fucked btw":** US House of Representatives, "The OPM Data Breach."

115 **At a talk one evening in Aspen:** For a full transcript of the panel, see "Beyond the Build: Leveraging the Cyber Mission Force," Aspen Security Forum, July 23, 2015, aspensecurityforum.org/wp-content/uploads/2015/07/Beyond -the-Build-Leveraging-the-Cyber-Mission-Force.pdf.

116 **"Protecting our federal employee data"**: "OPM to Notify Employees of Cybersecurity Incident," US Office of Personnel Management, June 4, 2015, www .opm.gov/news/releases/2015/06/opm-to-notify-employees-of-cybersecurity -incident/.

116 **"You have to kind of salute the Chinese for what they did"**: Damian Paletta, "U.S. Intelligence Chief James Clapper Suggests China Behind OPM Breach," *Wall Street Journal,* June 25, 2015, www.wsj.com/articles/ SB10007111583511843695404581069863170899504?

117 **"people who live in glass houses shouldn't throw rocks"**: "Cybersecurity Policy and Threats," Hearing Before the Senate Armed Services Committe, September 29, 2015, www.armed-services.senate.gov/imo/media/doc/15-75%20 -%209-29-15.pdf.

118 **some of its proprietary data had been stolen:** The deal was ill fated. In 2017 Westinghouse filed for bankruptcy, following billion-dollar cost overruns and massive delays on multiple nuclear reactor project sites ranging from China to the American South. To be clear, China didn't make Westinghouse fail. But the fall of Westinghouse due to flawed logistics and design glitches illustrates just how difficult it is to research and develop new products, particularly those on a massive scale. That is one step that China was eager to avoid by stealing the designs.

118 **There were other victims:** "Indictment Criminal No. 14-118," US District Court Western District of Pennsylvania, May 1, 2014, www.justice.gov/iso/ opa/resources/5122014519132358461949.pdf.

120 **the approval came:** "U.S. Charges Five Chinese Military Hackers for Cyber Espionage Against U.S. Corporations and a Labor Organization for Commercial Advantage," US Department of Justice, May 19, 2014, www.justice .gov/opa/pr/us-charges-five-chinese-military-hackers-cyber-espionage-against -us-corporations-and-labor.

121 **The Chinese were blindsided:** "China Reacts Strongly to US Announcement of Indictment Against Chinese Personnel," Ministry of Foreign Affairs of the People's Republic of China, May 20, 2014, www.fmprc.gov.cn/mfa_eng/xwfw _665399/s2510_665401/2535_665405/t1157520.shtml.

121 **Obama's team . . . promptly threatened to impose sanctions:** Julie Hirschfeld Davis, "Obama Hints at Sanctions Against China over Cyberattacks," *New York Times,* September 17, 2015, www.nytimes.com/2015/09/17/us/politics/ obama-hints-at-sanctions-against-china-over-cyberattacks.html.

122 **Susan Rice, Obama's national security adviser, was dispatched to Beijing:** David E. Sanger, "U.S. and China Seek Arms Deal for Cyberspace," *New York Times,* September 20, 2015, www.nytimes.com/2015/09/20/world/asia/us-and -china-seek-arms-deal-for-cyberspace.html.

123 **Obama told American business leaders:** Ibid.

124 **Obama had invited all the Silicon Valley royalty:** Gardiner Harris, "State

Dinner for Xi Jinping Has High-Tech Flavor," *New York Times,* September 26, 2015, www.nytimes.com/2015/09/26/world/asia/state-dinner-for-xi-jinping -has-high-tech-flavor.html.

124 **it seemed to work right away:** Most government officials believe there has been a significant drop-off in state-sponsored theft of intellectual property. But at a briefing for reporters in late 2017, CIA analysts declined to say whether they saw any improvement. By 2018, a number of experts believed that Chinese hacking was virtually undeterred. Others held that the Chinese simply changed tactics, investing in American technologies, as described in chapter 11. See David E. Sanger, "Chinese Curb Cyberattacks on U.S. Interests, Report Finds," *New York Times,* June 21, 2016, www.nytimes.com/2016/ 06/21/us/politics/china-us-cyber-spying.html.

CHAPTER VI: THE KIMS STRIKE BACK

125 **had already written a searing letter of protest:** "North Korea Complains to UN about Film Starring Rogen, Franco," Reuters, July 9, 2014, uk.reuters .com/article/uk-northkorea-un-film/north-korea-complains-to-un-about-film -starring-rogen-franco-idUKKBN0FE21B20140709.

125 **North Korea began issuing threats against the United States:** BBC News, "The Interview: A Guide to the Cyber Attack on Hollywood," BBC News, December 29, 2014, www.bbc.com/news/entertainment-arts-30512032.

127 **Kim Heung-kwang, a North Korean defector:** David E. Sanger and Martin Fackler, "N.S.A. Breached North Korean Networks Before Sony Attack, Officials Say," *New York Times,* January 19, 2015, www.nytimes.com/2015/ 01/19/world/asia/nsa-tapped-into-north-korean-networks-before-sony-attack -officials-say.html.

128 **Jang Sae-yul, a former North Korean army programmer:** Ibid.

128 **"If warfare was about bullets and oil until now":** David E. Sanger, David Kirkpatrick, and Nicole Perlroth, "The World Once Laughed at North Korean Cyberpower. No More," *New York Times,* October 16, 2017, www.nytimes .com/2017/10/15/world/asia/north-korea-hacking-cyber-sony.html.

129 **"There was an enormous growth in capability":** Ibid.

130 **"they have one of the most successful cyber programs":** Ibid.

134 **the project was canceled:** Ibid.

134 **began peppering Clapper with questions:** Siobhan Gorman and Adam Entous wrote the first full account of the visit: "U.S. Spy Chief Gives Inside Look at North Korea Prisoner Deal," *Wall Street Journal,* November 14, 2014, www .wsj.com/articles/u-s-spy-chief-gives-inside-look-at-north-korea-prisoner-deal -1416008783.

135 **"spent most of the meal berating me":** "Remarks as Delivered by the Honorable James R. Clapper Director of National Intelligence," Office of the Director of National Intelligence, January 7, 2015, www.dni.gov/index.php/

newsroom/speeches-interviews/speeches-interviews-2015/item/1156-remarks
-as-delivered-by-dni-james-r-clapper-on-national-intelligence-north-korea
-and-the-national-cyber-discussion-at-the-international-conference-on-cyber
-security.

136 **likely knew a lot about the Sony hack:** American intelligence officials wouldn't
conclude this until after the attack had happened.

136 **justified the United States' failure:** Rick Gladstone and David E. Sanger, "Security Council Tightens Economic Vise on North Korea, Blocking Fuel, Ships
and Workers," *New York Times,* December 23, 2017, www.nytimes.com/2017/
12/22/world/asia/north-korea-security-council-nuclear-missile-sanctions
.html.

140 **"minimally talented spoiled brat":** These details appeared in a story written
with my colleague Martin Fackler, who conducted some of the interviews with
defectors in Seoul. "N.S.A. Breached North Korean Networks Before Sony
Attack, Officials Say," *New York Times,* January 19, 2015, www.nytimes.com/
2015/01/19/world/asia/nsa-tapped-into-north-korean-networks-before-sony
-attack-officials-say.html.

140 **"Sony Pictures will be bombarded as a whole":** Michael Cieply and Brooks
Barnes, "Sony Cyberattack, First a Nuisance, Swiftly Grew into a Firestorm,"
New York Times, December 31, 2014, www.nytimes.com/2014/12/31/business/
media/sony-attack-first-a-nuisance-swiftly-grew-into-a-firestorm-.html.

140 **emails from Seth Rogen to studio executive:** Martin Fackler, Brooks Barnes,
and David E. Sanger, "Sony's International Incident: Making Kim Jong-un's
Head Explode," *New York Times,* December 15, 2014, www.nytimes.com/
2014/12/15/world/sonys-international-incident-making-kims-head-explode
.html.

143 **"cyber vandalism":** Eric Bradner, "Obama: North Korea's Hack Not War, but
'Cybervandalism,'" CNN, December 24, 2014, www.cnn.com/2014/12/21/
politics/obama-north-koreas-hack-not-war-but-cyber-vandalism/index.html.

143 **"the world will see what an awful movie":** Andrea Peterson, "Sony Pictures
Hackers Invoke 9/11 While Threatening Theaters That Show 'The Interview,'"
Washington Post, December 16, 2014, www.washingtonpost.com/news/the
-switch/wp/2014/12/16/sony-pictures-hackers-invoke-911-while-threatening
-theaters-that-show-the-interview/?utm_term=.b1ead7061843.

144 **took the unprecedented step of blaming North Korea:** David E. Sanger, Michael S. Schmidt, and Nicole Perlroth, "Obama Vows a Response to Cyberattack on Sony," *New York Times,* December 20, 2014, www.nytimes.com/
2014/12/20/world/fbi-accuses-north-korean-government-in-cyberattack-on
-sony-pictures.html.

145 **North Koreans had not expected the United States to conclude so quickly:**
Lewis laid out this argument in greater length in "North Korea and Cyber
Catastrophe—Don't Hold Your Breath," *38 North,* January 12, 2018, www
.38north.org/2018/01/jalewis011218/.

147 **presented a new cyber strategy at Stanford:** "Remarks by Secretary Carter at the Drell Lecture Cemex Auditorium," US Department of Defense, April 23, 2015, www.defense.gov/News/Transcripts/Transcript-View/Article/607043/ remarks-by-secretary-carter-at-the-drell-lecture-cemex-auditorium-stanford -grad/.

150 **the White House announced some weak economic sanctions:** Choe Sang-Hun, "North Korea Offers U.S. Deal to Halt Nuclear Test," *New York Times,* January 11, 2015, www.nytimes.com/2015/01/11/world/asia/north-korea -offers-us-deal-to-halt-nuclear-test-.html.

CHAPTER VII: PUTIN'S PETRI DISH

154 **relying on the advice and services of Paul Manafort:** Two years later, Manafort reemerged as chairman of Trump's 2016 campaign—and my one in-person encounter with him came when Maggie Haberman and I interviewed Trump at the Republican National Convention. After greeting us in Trump's hotel room, Manafort left before the inevitable Russia questions began.

154 **Putin's cyber army went to work:** Mark Clayton, "Ukraine Election Narrowly Avoided 'Wanton Destruction' from Hackers," *Christian Science Monitor,* June 17, 2014, www.csmonitor.com/World/Passcode/2014/0617/Ukraine -election-narrowly-avoided-wanton-destruction-from-hackers.

155 **announced the phony results:** Farangis Najibullah, "Russian TV Announces Right Sector Leader Led Ukraine Polls," RadioFreeEurope/RadioLiberty, May 26, 2014, www.rferl.org/a/russian-tv-announces-right-sector-leader -yarosh-led-ukraine-polls/25398882.html.

157 **The strategy was hardly a state secret:** "The Value of Science in Anticipating," *Military-Industrial Courier* [in Russian], February 26, 2013, www.vpk-news .ru/articles/14632.

157 **"What's new is not the basic model":** Joseph S. Nye Jr., "How Sharp Power Threatens Soft Power," *Foreign Affairs,* January 24, 2018, www.foreignaffairs .com/articles/china/2018-01-24/how-sharp-power-threatens-soft-power.

158 **There were many critiques of the Gerasimov doctrine:** A good example of such a critique is Michael Kofman and Matthew Rojansky, "A Closer Look at Russia's 'Hybrid War,'" Kennan Cable 7, The Wilson Center, April 2015, www .wilsoncenter.org/sites/default/files/7-KENNAN%20CABLE-ROJANSKY %20KOFMAN.pdf.

159 *We own you.*: Some of this chapter draws on a paper I wrote for the Aspen Strategy Group. See David E. Sanger, "Short of War: Cyber Conflict and the Corrosion of the International Order," in *The World Turned Upside Down: Maintaining American Leadership in a Dangerous Age,* Aspen Strategy Group, 2017.

159 **handed it over to the Ukrainians:** Ivo H. Daalder, "Responding to Russia's Resurgence: Not Quiet on the Eastern Front," *Foreign Affairs* 96, November/ December 2017, 30–38.

160 **Putin sought to boost the legitimacy of his actions:** David Adesnik, "How

Russia Rigged the Crimean Referendum," *Forbes,* March 18, 2014, www.forbes .com/sites/davidadesnik/2014/03/18/how-russia-rigged-crimean-referendum/ #774963966d41.

160 **the United States was oddly passive:** Jeffrey Goldberg, "The Obama Doctrine," *The Atlantic,* April 2016, www.theatlantic.com/magazine/archive/ 2016/04/the-obama-doctrine/471525.

161 **He did it in typical Trumpian style:** "Transcript: Donald Trump Expounds on His Foreign Policy Views," *New York Times,* March 27, 2016, www.nytimes .com/2016/03/27/us/politics/donald-trump-transcript.html.

162 **"Cyber Pearl Harbor":** Elisabeth Bumiller and Thom Shanker, "Panetta Warns of Dire Threat of Cyberattack on U.S," *New York Times,* October 12, 2012, www.nytimes.com/2012/10/12/world/panetta-warns-of-dire-threat-of -cyberattack.html.

163 **anything that went wrong:** Emily O. Goldman and Michael Warner, "Why a Digital Pearl Harbor Makes Sense . . . and Is Possible," in George Perkovich and Ariel E. Levite, eds., *Understanding Cyber Conflict: 14 Analogies* (Washington, DC: Georgetown University Press, 2017), 147–61.

164 **already inside the American electric grid:** Nicole Perlroth and David E. Sanger, "Cyberattacks Put Russian Fingers on the Switch at Power Plants, U.S. Says," *New York Times,* March 16, 2018, www.nytimes.com/2018/03/15/ us/politics/russia-cyberattacks.html.

165 **that distinction may mean little:** Martin C. Libicki, *Cyberspace in Peace and War* (Annapolis, MD: Naval Institute Press, 2016), 288.

166 **turned out the lights for only 225,000 customers:** Robert M. Lee, Michael J. Assante, and Tim Conway, "Analysis of the Cyber Attack on the Ukrainian Power Grid," SANS ICS and the Electricity Information Sharing and Analysis Center, March 18, 2016, ics.sans.org/media/E-ISAC_SANS_Ukraine_DUC _5.pdf.

168 **"the outages were caused by the use of the control systems":** Ibid., 3.

169 **government had been hit by 6,500 cyberattacks:** "Ukraine Power Cut 'Was Cyber-attack,'" BBC News, January 11, 2017, www.bbc.com/news/technology -38573074.

170 **Andy Greenberg of *Wired* wrote:** Andy Greenberg, "How an Entire Nation Became Russia's Test Lab for Cyberwar," *Wired,* June 20, 2017, www.wired.com/ story/russian-hackers-attack-ukraine/.

CHAPTER VIII: THE FUMBLE

171 **"I cannot forecast to you the action of Russia":** Alan Cowell, "Churchill's Definition of Russia Still Rings True," *New York Times,* August 1, 2008, www .nytimes.com/2008/08/01/world/europe/01iht-letter.1.14939466.html.

172 **suspicious Russian intrusion into the computer networks:** Eric Lipton, David E. Sanger, and Scott Shane, "The Perfect Weapon: How Russian Cyberpower

Invaded the U.S.," *New York Times*, December 14, 2016, www.nytimes.com/ 2016/12/13/us/politics/russia-hack-election-dnc.html.

175 **Golos, the only independent election monitoring group:** " 'Hacking Attacks' Hit Russian Political Sites," BBC News, BBC, March 8, 2012, www.bbc.com/ news/technology-16032402.

176 **"She set the tone for some actors in our country":** David M. Herszenhorn and Ellen Barry, "Putin Contends Clinton Incited Unrest over Vote," *New York Times*, December 9, 2011, www.nytimes.com/2011/12/09/world/europe/putin -accuses-clinton-of-instigating-russian-protests.html?mcubz=2.

176 **The United States did not exactly have clean hands:** Two informative sources on this history are Evan Osnos, David Remnick, and Joshua Yaffa, "Trump, Putin, and the New Cold War," *New Yorker*, March 6, 2017, www .newyorker.com/magazine/2017/03/06/trump-putin-and-the-new-cold-war; and Calder Walton, " 'Active Measures': A History of Russian Interference in US Elections," *Prospect*, December 23, 2016, www.prospectmagazine.co.uk/ science-and-technology/active-measures-a-history-of-russian-interference-in -us-elections.

177 **Putin's moral equivalence:** As Jackson Diehl of the *Washington Post* would later suggest of the 2016 US election: "Putin developed an obsession with 'color revolutions,' which he is convinced are neither spontaneous nor locally organized, but orchestrated by the United States . . . Putin is trying to deliver to the American political elite what he believes is a dose of its own medicine. He is attempting to ignite—with the help, unwitting or otherwise, of Donald Trump—a U.S. color revolution." See "Putin's Hope to Ignite a Eurasia-Style Protest in the United States," October 16, 2016, www.washingtonpost.com/ opinions/global-opinions/putins-hope-to-ignite-a-eurasia-style-protest-in-the -united-states/2016/10/16/0f271a60-90a4-11e6-9c85-ac42097b8cc0_story .html?utm_term=.f8bb8e047e48.

177 **in the bad old days:** Scott Shane, "Russia Isn't the Only One Meddling in Elections. We Do It, Too," *New York Times*, February 18, 2018, www .nytimes.com/2018/02/17/sunday-review/russia-isnt-the-only-one-meddling -in-elections-we-do-it-too.html.

177 **Patient Zero:** Susan B. Glasser, "Victoria Nuland: The Full Transcript," *Politico*, February 5, 2018, www.politico.com/magazine/story/2018/02/05/victoria -nuland-the-full-transcript-216936.

180 **the audio, edited, suddenly appeared on YouTube:** "US Blames Russia for Leak of Undiplomatic Language from Top Official," *Guardian*, February 6, 2014, www.theguardian.com/world/2014/feb/06/us-russia-eu-victoria-nuland.

182 **the man who would work to alter the 2016 election:** Neil MacFarquhar, "Yevgeny Prigozhin, Russian Oligarch Indicted by U.S., Is Known as 'Putin's Cook,' " *New York Times*, February 17, 2018, www.nytimes.com/2018/02/16/ world/europe/prigozhin-russia-indictment-mueller.html.

183 **social media could just as easily incite disagreements:** Adrian Chen, "The Real Paranoia-Inducing Purpose of Russian Hacks," *New Yorker,* July 27, 2016, www.newyorker.com/news/news-desk/the-real-paranoia-inducing-purpose-of -russian-hacks.

184 **rose pretty fast:** "United States of America v. Internet Research Agency," indictment of the 13 Internet Research Agency members, filed by Special Counsel Robert Mueller, February 16, 2018, www.justice.gov/file/1035477/ download.

184 **took the propaganda battle to the enemy's territory:** Adrian Chen, "The Agency," *New York Times,* June 7, 2015, www.nytimes.com/2015/06/07/ magazine/the-agency.html.

184 **learned to "troll" critics of Putin and journalists:** Alexis C. Madrigal, "Russia's Troll Operation Was Not That Sophisticated," *The Atlantic,* February 19, 2018, www.theatlantic.com/technology/archive/2018/02/the-russian -conspiracy-to-commit-audience-development/553685.

185 **Then they moved on to advertising:** Scott Shane, "The Fake Americans Russia Created to Influence the Election," *New York Times,* September 8, 2017, www.nytimes.com/2017/09/07/us/politics/russia-facebook-twitter-election .html.

185 **Putin's trolls posed as Americans:** April Glaser, "What We Know About How Russia's Internet Research Agency Meddled in the 2016 Election," *Slate,* February 16, 2018, slate.com/technology/2018/02/what-we-know-about-the -internet-research-agency-and-how-it-meddled-in-the-2016-election.html.

185 **Ryan Lizza's reporting in the *New Yorker*:** Ryan Lizza, "How Trump Helps Russian Trolls," *New Yorker,* November 2, 2017, www.newyorker.com/news/ our-columnists/how-trump-helps-russian-trolls.

185 **The agency dispatched two of their experts:** Ivan Nechepurenko and Michael Schwirtz, "The Troll Farm: What We Know About 13 Russians Indicted by the U.S.," *New York Times,* February 17, 2018, www.nytimes.com/2018/02/ 17/world/europe/russians-indicted-mueller.html.

186 **the trolls tested out a new tactic:** Editorial Staff, "How the 'Troll Factory' Worked in the US Elections," RBC (in Russian), October 17, 2017, www.rbc .ru/magazine/2017/11/59e0c17d9a79470e05a9e6c1.

186 **The magazine reported:** Hannah Levintova, "Russian Journalists Just Published a Bombshell Investigation About a Kremlin-Linked 'Troll Factory,'" *Mother Jones,* October 18, 2017, www.motherjones.com/politics/2017/10/ russian-journalists-just-published-a-bombshell-investigation-about-a-kremlin -linked-troll-factory.

187 **Kevin Mandia and his firm's:** Relevant reports by Fireye, Mandia's firm, viewable at www.fireeye.com/content/dam/fireeye-www/solutions/pdfs/st-senate -intel-committee-russia-election.pdfl and www.fireeye.com/blog/threat

-research/2014/10/apt28-a-window-into-russias-cyber-espionage-operations
.html.

189 **their White House tour:** Michael S. Schmidt and David E. Sanger, "Russian Hackers Read Obama's Unclassified Emails, Officials Say," *New York Times,* April 26, 2015, www.nytimes.com/2015/04/26/us/russian-hackers-read-obamas-unclassified-emails-officials-say.html.

189 **"hand-to-hand combat":** Joseph Marks, "NSA Engaged in Massive Battle with Russian Hackers in 2014," Nextgov, April 3, 2017, www.nextgov.com/cybersecurity/2017/04/nsa-engaged-massive-battle-russian-hackers-2014/136683/.

192 **Tamene and his colleagues had met the FBI:** Eric Lipton, David E. Sanger, and Scott Shane, "The Perfect Weapon: How Russian Cyberpower Invaded the U.S."

CHAPTER IX: WARNING FROM THE COTSWOLDS

196 **The Doughnut's design was very Silicon Valley:** In one of those great transcontinental reversals, the Doughnut helped inspire Apple's design for its own headquarters, a bit more than five thousand miles away.

198 **the "Five Eyes":** For a good discussion of Five Eyes, see Levi Maxey, "Five Eyes Intel Sharing Unhindered by Trump Tweets," *The Cipher Brief,* February 20, 2018, www.thecipherbrief.com/five-eyes-intel-sharing-unhindered-trump-tweets.

200 **"summer project for Menwith":** The best description of this operation appeared in the British newspaper *The Guardian,* which also published many of the Snowden papers. See Ewen MacAskill, Julian Borger, Nick Hopkins, Nick Davies, and James Ball, "GCHQ Taps Fibre-Optic Cables for Secret Access to World's Communications," June 21, 2013, www.theguardian.com/uk/2013/jun/21/gchq-cables-secret-world-communications-nsa.

202 **they worked twelve-hour shifts:** Neil MacFarquhar, "Inside the Russian Troll Factory: Zombies and a Breakneck Pace," *New York Times,* February 19, 2018, www.nytimes.com/2018/02/18/world/europe/russia-troll-factory.html.

204 **had been hacked by not one Russian intelligence group but two:** Dmitri Alperovitch, "Bears in the Midst: Intrusion into the Democratic National Committee," *CrowdStrike,* June 15, 2016, www.crowdstrike.com/blog/bears-midst-intrusion-democratic-national-committee/.

204 **And both had left plenty of fingerprints:** Eric Lipton, David E. Sanger, and Scott Shane, "The Perfect Weapon: How Russian Cyberpower Invaded the U.S."

207 **The day after the *Post* and the *Times* ran their stories:** Ellen Nakashima, "Russian Government Hackers Penetrated DNC, Stole Opposition Research on Trump," *Washington Post,* June 15, 2016, www.washingtonpost.com/world/national-security/russian-government-hackers-penetrated-dnc

-stole-opposition-research-on-trump/2016/06/14/cf006cb4-316e-11e6-8ff7
-7b6c1998b7a0_story.html?; David E. Sanger and Nick Corasaniti, "D.N.C.
Says Russian Hackers Penetrated Its Files, Including Dossier on Donald
Trump," *New York Times,* June 14, 2016, www.nytimes.com/2016/06/15/us/
politics/russian-hackers-dnc-trump.html.

207 **A persona with the screen name Guccifer 2.0:** Rob Price, "RESEARCH-
ERS: Yes, Russia Really Did Hack the Democratic National Congress," *Business Insider Australia,* June 21, 2016, www.businessinsider.com.au/security
-researchers-russian-spies-hacked-dnc-guccifer-2-possible-disinformation
-campaign-2016-6.

208 **said he was Romanian:** Lorenzo Franceschi-Bicchierai, "Alleged Russian
Hacker 'Guccifer 2.0' Is Back After Months Of Silence," *Vice,* January 12,
2017, motherboard.vice.com/en_us/article/9a3m7p/alleged-russian-hacker
-guccifer-20-is-back-after-months-of-silence. The transcript of Lorenzo's interview with Guccifer 2.0 is viewable at motherboard.vice.com/en_us/article/
yp3bbv/dnc-hacker-guccifer-20-full-interview-transcript.

210 **our second foreign-policy interview with Trump:** "Transcript: Donald Trump
Expounds on His Foreign Policy Views," *New York Times,* March 26, 2016,
www.nytimes.com/2016/03/27/us/politics/donald-trump-transcript.html.

213 **Nicole Perlroth and I wrote:** David E. Sanger and Nicole Perlroth, "As Democrats Gather, a Russian Subplot Raises Intrigue," *New York Times,* July 25, 2016,
www.nytimes.com/2016/07/25/us/politics/donald-trump-russia-emails.html.

CHAPTER X: THE SLOW AWAKENING

215 **"hallmark of our democracy":** "Obama's Last News Conference: Full Transcript and Video," *New York Times,* January 18, 2017, www.nytimes.com/
2017/01/18/us/politics/obama-final-press-conference.html.

218 **"high confidence":** David E. Sanger and Scott Shane, "Russian Hackers Acted
to Aid Trump in Election, U.S. Says," *New York Times,* December 10, 2016,
www.nytimes.com/2016/12/09/us/obama-russia-election-hack.html.

218 **Brennan later argued:** Erika Fry, "Ex-CIA Director: Russia Wanted Hillary Clinton 'Bloodied' By Her Inauguration," *Fortune,* July 19, 2017, fortune
.com/2017/07/19/cia-director-russia-hillary-clinton/.

219 **had similar vulnerabilities:** David E. Sanger and Charlie Savage. "Sowing
Doubt Is Seen as Prime Danger in Hacking Voting System," *New York Times,*
September 15, 2016, www.nytimes.com/2016/09/15/us/politics/sowing-doubt
-is-seen-as-prime-danger-in-hacking-voting-system.html.

219 **Fox News on August 1:** David Weigel, "For Trump, a New 'Rigged' System:
The Election Itself," *Washington Post,* August 2, 2016, www.washingtonpost
.com/politics/for-trump-a-new-rigged-system-the-election-itself/2016/08/02/
d9fb33b0-58c4-11e6-9aee-8075993d73a2_story.html?

220 **"voter fraud is all too common"**: Jeremy Diamond, "Trump: 'I'm Afraid the Election's Going to Be Rigged,'" CNN, August 2, 2016, www.cnn.com/2016/08/01/politics/donald-trump-election-2016-rigged/index.html.

220 **"critical infrastructure"**: Julie Hirschfeld Davis, "U.S. Seeks to Protect Voting System From Cyberattacks," *New York Times,* August 4, 2016, www.nytimes.com/2016/08/04/us/politics/us-seeks-to-protect-voting-system-against-cyberattacks.html.

220 **"troubling reports"**: Erica R. Hendry, "Read Jeh Johnson's Prepared Testimony on Russia," PBS, June 20, 2017, www.pbs.org/newshour/politics/read-jeh-johnsons-prepared-testimony-russia.

221 **state-run election systems:** David E. Sanger and Charlie Savage, "U.S. Says Russia Directed Hacks to Influence Elections," *New York Times,* October 8, 2016, www.nytimes.com/2016/10/08/us/politics/us-formally-accuses-russia-of-stealing-dnc-emails.html.

223 **first rule of foreign policy:** Mark Landler, "In Obama's Speeches, a Shifting Tone on Terror," *New York Times,* June 1, 2014, www.nytimes.com/2014/06/01/world/americas/in-obamas-speeches-a-shifting-tone-on-terror.html.

224 **session with twelve congressional leaders:** Susan B. Glasser, "Did Obama Blow It on the Russian Hacking?" *Politico,* April 3, 2017, www.politico.eu/article/did-obama-blow-it-on-the-russian-hacking-us-elections-vladimir-putin-donald-trump-lisa-monaco/.

225 **interceded in the election:** David E. Sanger, "What Is Russia Up To, and Is It Time to Draw the Line?" *New York Times,* September 30, 2016, www.nytimes.com/2016/09/30/world/europe/for-veterans-of-the-cold-war-a-hostile-russia-feels-familiar.html.

225 **"they have the cyber tools":** Ibid.

225 **envelop its debates in great secrecy:** Some of the best details of the debates appeared in a June 23, 2017, reconstruction of events in the *Washington Post* by Greg Miller, Ellen Nakashima, and Adam Entous: "Obama's Secret Struggle to Punish Russia for Putin's Election Assault," www.washingtonpost.com/graphics/2017/world/national-security/obama-putin-election-hacking/?utm_term=.92aacc38a2da.

226 **Something else had sunk in:** This investigation into the Shadow Brokers is drawn from extensive reporting that my colleagues Nicole Perlroth and Scott Shane and I did and published in a November 12, 2017, *Times* feature, "Security Breach and Spilled Secrets Have Shaken the N.S.A. to Its Core."

229 **Pho had apparently brought home:** In other words, Kaspersky acknowledged finding the NSA's software on one of its customers' computers and removing it, but insists it was destroyed; the United States believes it was passed to Russian intelligence. That is why the United States banned Kaspersky products from any government computers in 2017. Pho was secretly arrested in 2015, but the case became public only in December 2017, when he pled guilty to a single count of "willful retention of national defense information."

231 **"somebody sitting on their bed"**: Aaron Blake, "The First Trump-Clinton Presidential Debate Transcript, Annotated," *Washington Post,* September 26, 2016, www.washingtonpost.com/news/the-fix/wp/2016/09/26/the-first-trump -clinton-presidential-debate-transcript-annotated/?

231 **Office of the Director of National Intelligence:** www.dhs.gov/news/2016/10/ 07/joint-statement-department-homeland-security-and-office-director -national.

232 **"We believe, based on the scope":** Kate Conger, "U.S. Officially Attributes DNC Hack to Russia," *TechCrunch,* October 7, 2016, techcrunch.com/2016/ 10/07/u-s-attributes-dnc-hack-russia/.

232 *Access Hollywood*: David A. Fahrenthold, "Trump Recorded Having Extremely Lewd Conversation About Women in 2005," *Washington Post,* October 8, 2016, www.washingtonpost.com/politics/trump-recorded-having -extremely-lewd-conversation-about-women-in-2005/2016/10/07/3b9ce776 -8cb4-11e6-bf8a-3d26847eeed4_story.html?utm_term=.302520d75fcb.

233 **shortcomings as a candidate:** Amy Chozick, Nicholas Confessore, and Michael Barbaro, "Leaked Speech Excerpts Show a Hillary Clinton at Ease with Wall Street," *New York Times,* October 8, 2016, www.nytimes.com/2016/ 10/08/us/politics/hillary-clinton-speeches-wikileaks.html.

234 **"process of voting":** Politico Staff, "Full Transcript: President Obama's Final End-of-Year Press Conference," *Politico,* December 16, 2016, www.politico .com/story/2016/12/obama-press-conference-transcript-232763.

234 **hacking into American infrastructure:** David E. Sanger, "Obama Strikes Back at Russia for Election Hacking," *New York Times,* December 30, 2016, www.nytimes.com/2016/12/29/us/politics/russia-election-hacking-sanctions .html.

234 **black wisps of smoke:** "Black Smoke Pours from Chimney at Russian Consulate in San Francisco," CBS News, September 1, 2017, www.cbsnews.com/ news/black-smoke-chimney-russian-consulate-san-francisco/.

236 **sat down in Hamburg, Germany:** Julie Hirschfeld Davis, David E. Sanger, and Glenn Thrush, "Trump Questions Putin on Election Meddling at Eagerly Awaited Encounter," *New York Times,* July 8, 2017, www.nytimes.com/2017/ 07/07/world/europe/trump-putin-g20.html.

237 **the initial mistakes:** Eric Lipton, David E. Sanger, and Scott Shane, "The Perfect Weapon: How Russian Cyberpower Invaded the U.S.," *New York Times,* December 13, 2016, www.nytimes.com/2016/12/13/us/politics/russia-hack -election-dnc.html.

CHAPTER XI: THREE CRISES IN THE VALLEY

240 **"If you had asked me":** Kevin Roose and Sheera Frenkel, "Mark Zuckerberg's Reckoning: 'This Is a Major Trust Issue,'" *New York Times,* March 22, 2018, www.nytimes.com/2018/03/21/technology/mark-zuckerberg-q-and-a.html.

240 **Twenty minutes after:** "Paris Attacks: What Happened on the Night," BBC News, December 9, 2015, www.bbc.com/news/world-europe-34818994.

241 **"be unforgiving with the barbarians from Daesh":** Adam Nossiter, Aurelien Breeden, and Katrin Bennhold, "Three Teams of Coordinated Attackers Carried Out Assault on Paris, Officials Say; Hollande Blames ISIS," *New York Times,* November 15, 2015, www.nytimes.com/2015/11/15/world/europe/paris-terrorist-attacks.html.

243 **"It's almost never as cool":** David E. Sanger and Eric Schmitt, "U.S. Cyberweapons, Used Against Iran and North Korea, Are a Disappointment Against ISIS," *New York Times,* June 13, 2017, www.nytimes.com/2017/06/12/world/middleeast/isis-cyber.html.

244 **"We are dropping cyber bombs":** David E. Sanger, "U.S. Cyberattacks Target ISIS in a New Line of Combat," *New York Times,* April 25, 2016, www.nytimes.com/2016/04/25/us/politics/us-directs-cyberweapons-at-isis-for-first-time.html?

244 **"our cyber operations are disrupting their command-and-control":** The White House, Office of the Press Secretary, "Statement by the President on Progress in the Fight Against ISIL," April 13, 2016, obamawhitehouse.archives.gov/the-press-office/2016/04/13/statement-president-progress-fight-against-isil.

246 **three months behind schedule:** Ellen Nakashima, "U.S. Military Cyber Operation to Attack ISIS Last Year Sparked Heated Debate over Alerting Allies," *Washington Post,* May 9, 2017, www.washingtonpost.com/world/national-security/us-military-cyber-operation-to-attack-isis-last-year-sparked-heated-debate-over-alerting-allies/2017/05/08/93a120a2-30d5-11e7-9dec-764dc781686f_story.html?

247 **Carter wrote a blistering assessment:** Ash Carter, *A Lasting Defeat: The Campaign to Destroy ISIS,* Belfer Center for Science and International Affairs, October 2017, www.belfercenter.org/LastingDefeat#6.

249 **Stamos, then the chief security officer at Yahoo!:** CNBC, "Yahoo Security Officer Confronts NSA Director," YouTube video, 0:20, February 28, 2015, accessed April 10, 2018, www.youtube.com/watch?v=jJZNvEPyjlw.

250 **"Facebook can't stop monetizing our personal data":** Kevin Roose, "Can Social Media Be Saved?," *New York Times,* March 29, 2018, www.nytimes.com/2018/03/28/technology/social-media-privacy.html.

251 *Aftenposten* **called the company out:** Espen Egil Hansen, "Dear Mark. I Am Writing This to Inform You That I Shall Not Comply with Your Requirement to Remove This Picture," *Aftenposten,* September 8, 2016, www.aftenposten.no/meninger/kommentar/i/G892Q/Dear-Mark-I-am-writing-this-to-inform-you-that-I-shall-not-comply-with-your-requirement-to-remove-this-picture.

252 **"Let's say that somebody uploads an ISIS formal propaganda video":** Monika Bickert, interview by Steve Inskeep, "How Facebook Uses Technology to

Block Terrorist-Related Content," *NPR Morning Edition,* June 22, 2017, www
.npr.org/sections/alltechconsidered/2017/06/22/533855547/how-facebook
-uses-technology-to-block-terrorist-related-content.

252 **Google, meanwhile, tried a different approach:** The YouTube Team, "Bringing New Redirect Method Features to YouTube," Official YouTube Blog,
July   20,   2017,   youtube.googleblog.com/2017/07/bringing-new-redirect
-method-features.html.

253 **"Personally I think the idea that fake news on Facebook":** Mark Zuckerberg, interview by David Kirkpatrick, "In Conversation with Mark Zuckerberg," *Techonomy,* November 17, 2016, techonomy.com/conf/te16/videos
-conversations-with-2/in-conversation-with-mark-zuckerberg/.

253 **The president took him into a private room:** Adam Entous, Elizabeth
Dwoskin, and Craig Timberg, "Obama Tried to Give Zuckerberg a Wake-
Up Call over Fake News on Facebook," *Washington Post,* September 24,
2017, www.washingtonpost.com/business/economy/obama-tried-to
-give-zuckerberg-a-wake-up-call-over-fake-news-on-facebook/2017/09/24/
15d19b12-ddac-4ad5-ac6e-ef909e1c1284_story.html?

254 **His study was delivered to the company's leadership:** Jen Weedon, William Nuland, and Alex Stamos, *Information Operations and Facebook,* Facebook, April 27, 2017, fbnewsroomus.files.wordpress.com/2017/04/facebook
-and-information-operations-v1.pdf.

255 **"'I must say, I don't think you get it'":** Brett Samuels, "Feinstein to Tech
Execs: 'I Don't Think You Get It,'" *The Hill,* November 1, 2017, thehill
.com/business-a-lobbying/358232-feinstein-to-tech-cos-i-dont-think-you
-get-it.

255 **"We have a responsibility":** Daniel Politi, "Facebook's Zuckerberg Takes Out
Full Page Ads to Say 'Sorry' for 'Breach of Trust,'" *Slate,* March 25, 2018, slate
.com/news-and-politics/2018/03/facebooks-zuckerberg-takes-out-full-page
-ads-to-say-sorry-for-breach-of-trust.html.

260 **those satellites was estimated to cost more than $94 billion:** David E.
Sanger and William Broad, "Tiny Satellites from Silicon Valley May Help
Track North Korea Missiles," *New York Times,* July 7, 2017, www.nytimes
.com/2017/07/06/world/asia/pentagon-spy-satellites-north-korea-missiles
.html.

261 **the Chinese spend 1.7 billion hours a day on Tencent apps:** Brad Stone and
Lulu Yilun Chen, "Tencent Dominates in China. Next Challenge Is Rest of
the World," Bloomberg, June 28, 2017, www.bloomberg.com/news/features/
2017-06-28/tencent-rules-china-the-problem-is-the-rest-of-the-world.

262 **The DIUx report's findings:** Early copies of the report circulated broadly and
eventually made their way onto the Internet. The Pentagon posted a version
in March 2018, stripped of the authors recommendations, on the DIUx website: www.diux.mil/, and Michael Brown and Pavneet Singh, *China's Technology Transfer Strategy: How Chinese Investments in Emerging Technology Enable*

a Strategic Competitor to Access the Crown Jewels of U.S. Innovation, January 2018, www.DIUx.mil.

263 **they did eighty-one deals in American artificial-intelligence companies:** Ibid.

CHAPTER XII: LEFT OF LAUNCH

268 **North Korea's missiles started falling out of the sky:** David E. Sanger and William J. Broad, "How U.S. Intelligence Agencies Underestimated North Korea," *New York Times,* January 7, 2018, www.nytimes.com/2018/01/06/world/asia/north-korea-nuclear-missile-intelligence.html.

269 **Kim Jong-un had ordered an investigation:** David E. Sanger and William J. Broad, "Hand of U.S. Leaves North Korea's Missile Program Shaken," *New York Times.* April 19, 2017, www.nytimes.com/2017/04/18/world/asia/north -korea-missile-program-sabotage.html.

270 **"It was a fiery, catastrophic attempt":** Foster Klug and Hyung-Jin Kim, "US: North Korean Missile Launch a 'Catastrophic' Failure," April 16, 2016, apnews.com/67c278f79593454e868ff3f707606ef3/seoul-says-north-korean -missile-launch-apparently-fails.

270 **"North Korea did not pose a threat to North America":** "Pentagon Spokes-man Comments on North Korean Missile Launch," US Northern Com-mand, July 28, 2017, www.northcom.mil/Newsroom/Article/1456396/pentagon-spokesman-comments-on-north-korean-missile-launch/.

273 **"We have never heard of him killing scientists":** Choe Sang-Hun, Motoko Rich, Natalie Reneau, and Audrey Carlsen, "Rocket Men: The Team Build-ing North Korea's Nuclear Missile," *New York Times,* December 15, 2017, www.nytimes.com/interactive/2017/12/15/world/asia/north-korea-scientists -weapons.html.

275 **Sixty years and more than $300 billion later:** David E. Sanger and William J. Broad, "Trump Inherits a Secret Cyberwar Against North Korean Missiles," *New York Times,* March 5, 2017, www.nytimes.com/2017/03/04/world/asia/north-korea-missile-program-sabotage.html.

275 **roughly 50 percent:** David E. Sanger and William J. Broad. "Downing North Korean Missiles Is Hard. So the U.S. Is Experimenting," *New York Times,* November 17, 2017, www.nytimes.com/2017/11/16/us/politics/north-korea -missile-defense-cyber-drones.html.

276 **Dempsey publicly announced a new "left of launch" effort:** "Joint Inte-grated Air and Missile Defense: Vision 2020," United States Joint Chiefs of Staff, December 5, 2013, www.jcs.mil/Portals/36/Documents/Publications/JointIAMDVision2020.pdf.

277 **a tiny glimpse of the effort came from Oren J. Falkowitz:** Nicole Perlroth, "The Chinese Hackers in the Back Office," *New York Times.* June 12, 2016, www.nytimes.com/2016/06/12/technology/the-chinese-hackers-in-the-back -office.html.

278 **a gathering of top antimissile experts:** Center for Strategic and International Studies, "Full Spectrum Missile Defense," December 4, 2015, www.csis.org/events/full-spectrum-missile-defense.

278 **"left of launch" strikes as "game changing":** William J. Broad and David E. Sanger, "U.S. Strategy to Hobble North Korea Was Hidden in Plain Sight," *New York Times,* March 4, 2017, www.nytimes.com/2017/03/04/world/asia/left-of-launch-missile-defense.html.

278 **how the United States would justify:** Ibid.

280 **America was blowing its lead:** The full transcript of the March 2016 interview is available at "Transcript: Donald Trump Expounds on His Foreign Policy Views," *New York Times,* March 26, 2016, www.nytimes.com/2016/03/27/us/politics/donald-trump-transcript.html.

281 **telling the government what we were preparing to publish:** Later, others from agencies that specialized in offensive cyber operations got involved.

284 **"the final stage in preparations":** Choe Sang-Hun, "Kim Jong-un Says North Korea Is Preparing to Test Long-Range Missile," *New York Times,* January 2, 2017, www.nytimes.com/2017/01/01/world/asia/north-korea-intercontinental-ballistic-missile-test-kim-jong-un.html.

284 **"It won't happen":** Donald J. Trump, "North Korea Just Stated That It Is in the Final Stages of Developing a Nuclear Weapon Capable of Reaching Parts of the U.S. It Won't Happen!" Twitter, January 2, 2017, twitter.com/realdonaldtrump/status/816057920223846400.

285 **steal $1 billion from the Bangladesh Central Bank:** David E. Sanger, David D. Kirkpatrick, and Nicole Perlroth, "The World Once Laughed at North Korean Cyberpower. No More." *New York Times,* October 16, 2017, www.nytimes.com/2017/10/15/world/asia/north-korea-hacking-cyber-sony.html.

287 **swept up 182 gigabytes of data:** Choe Sang-Hun, "North Korean Hackers Stole U.S.-South Korean Military Plans, Lawmaker Says," *New York Times,* October 11, 2017, www.nytimes.com/2017/10/10/world/asia/north-korea-hack-war-plans.html?

288 **NSA had warned the company:** Nicole Perlroth and David E. Sanger, "Hackers Hit Dozens of Countries Exploiting Stolen N.S.A. Tool," *New York Times,* May 13, 2017, www.nytimes.com/2017/05/12/world/europe/uk-national-health-service-cyberattack.html.

290 **Hutchins was later arrested in Las Vegas:** Selena Larson, "WannaCry 'Hero' Arrested for Creating Other Malware," CNNMoney, August 3, 2017, money.cnn.com/2017/08/03/technology/culture/malwaretech-arrested-las-vegas-trojan/index.html.

290 **declare that Kim Jong-un's government was responsible for Wanna-Cry:** "Press Briefing on the Attribution of the WannaCry Malware Attack to North Korea," The White House, December 19, 2017, www.whitehouse

.gov/briefings-statements/press-briefing-on-the-attribution-of-the-wannacry
-malware-attack-to-north-korea-121917/.

290 **Bossert was honest:** Ibid.

292 **among the worst hit:** Charlie Osborne, "NotPetya Ransomware Forced
Maersk to Reinstall 4000 Servers, 45000 PCs," *ZDNet,* January 26, 2018,
www.zdnet.com/article/maersk-forced-to-reinstall-4000-servers-45000-pcs
-due-to-notpetya-attack/.

292 **only one failed:** Sanger and Broad, "How U.S. Intelligence Agencies Underestimated North Korea."

293 **$4 billion in emergency funds:** Sanger and Broad, "Downing North Korean
Missiles Is Hard."

AFTERWORD

296 **I keep on my desk:** R. P. Hearne, *Airships in Peace and War,* London: John
Lane, the Bodley Head, 2nd edition, 1910.

298 **"Right now, if you look":** Stanford University address by Michael Rogers, November 3, 2014, www.nsa.gov/news-features/speeches-testimonies/speeches/
stanford.shtml.

299 **"we might not be so lucky":** The story is told in William J. Perry, *My Journey
to the Nuclear Brink* (Redwood City, CA: Stanford University Press, 2015).

301 **The approach Cyber Command described:** Brad Smith, *Archive and Maintain Cyberspace Superiority: Command Vision for US Cyber Command,*
https://assets.documentcloud.org/documents/4419681/Command-Vision-for-
USCYBERCOM-23-Mar-18.pdf.

306 **"Digital Geneva Convention":** Brad Smith, "The Need for a Digital Geneva
Convention," *Microsoft on the Issues,* March 9, 2017, blogs.microsoft.com/on
-the-issues/2017/02/14/need-digital-geneva-convention/.

INDEX